"Simply yet crisply written, [Miyoko] Chu's work summarizes much of what has recently been learned about the multiple lives of songbirds . . . Like John McPhee in his travels with geologists, Chu turns to the ornithological detectives to piece together the story . . . Chu's researchers have truly great tales to tell."
—*Washington Post*

"Chu covers theories and research with affection and passion . . . Whether or not you're aware of songbirds' exceptional lives of seasonal travel, Chu's book will both enlighten and delight."
—*Seattle Times*

"*Songbird Journeys* is a fascinating introduction to bird migration." —*Pittsburgh Post Gazette*

"Readers will be struck by the magic of migration, presented in this book in fascinating detail yet in a concise, readable format. The book also includes such a wide variety of detailed scientific work that even the most experienced birders will discover new facts about migratory birds." —*WildBird*

"For those of us excited by the mysteries of migration and potential research adventures, *Songbird Journeys* serves as a useful guide. Travel instructions encourage birders to visit a few of the best locations from Alberta to Panama for observing migratory birds throughout the year." —*Birder's World*

"A unique and delightful account of the migration ecology of songbirds in North America. This book will be of greatest interest to birders, but is also an informative good read for anyone interested in migratory birds." —*Journal of Field Ornithology*

"By mixing crisp writing with patient explanation, Miyoko Chu has produced something rare—a book that offers something for both expert ornithologists and backyard bird lovers."
—Mark Obmascik, author of *The Big Year*

"Miyoko Chu captures both the sweep and the drama of bird migration. *Songbird Journeys* mingles passion, beauty and science into a surprising, fascinating whole."
—Scott Weidensaul, author of *Living on the Wind*

"*Songbird Journeys* is filled with solid and current scientific information while beautifully expressing what it is about birds that touches our minds and hearts."
—Laura Erickson, author of *101 Ways to Help Birds*

SONGBIRD JOURNEYS

*Four Seasons in the Lives
of Migratory Birds*

MIYOKO CHU

Walker & Company
New York

Published by Walker Publishing Company, Inc., New York
Distributed to the trade by Holtzbrinck Publishers

All papers used by Walker & Company are natural, recyclable products made from wood grown in well-managed forests. The manufacturing processes conform to the environmental regulations of the country of origin.

THE LIBRARY OF CONGRESS HAS CATAOLGED THE HARDCOVER EDITION UNDER LCCN: 2006278075

First published in the United States by Walker & Company in 2006
This paperback edition published in 2007

Paperback ISBN-10: 0-8027-1518-4
ISBN-13: 978-0-8027-1518-0

Visit Walker & Company's Web site at www.walkerbooks.com

10 9 8 7 6 5 4 3 2 1

Typeset by Westchester Book Group
Printed in the United States of America by Quebecor World Fairfield

For Mark Chao
and for our children,
Francesca and Tilden

CONTENTS

WINTER

*I*NTRODUCTION

Summer and Autumn, Winter, Spring,
*   Each season of the varied year,*
Doth each for us a lesson bring,
*   If we but turn the listening ear.*
— *Jones Very, from "Nature Intelligible"*

EVERY SPRING, an extraordinary wildlife migration passes unseen within thousands of feet of our own neighborhoods. Millions of songbirds stream overhead at night, sometimes so closely that you can hear their faint, high-pitched calls as they pass by in the dark. As you sleep, a thousand orioles might travel through a slice of sky above your bed, northward bound after a winter in Mexico. Mixed in among them, multitudes of tanagers from tropical evergreen forests and buntings from the Caribbean head for the temperate woodlands where they will raise their young. Thousands of warblers flit past throughout the night, on their way from the tropics to the boreal forests of Canada.

Bobolinks ride the winds from the pampas of Argentina to the open grasslands of North America, while thrushes from Central America wing their way toward the broad taiga of Alaska and Canada.

Before daybreak, the birds sift down from the sky and settle into the trees. After a long night of travel, the birds begin singing at dawn as energetically as if they have just woken from a restful sleep. In the woods all day, songbirds sing and search for food from the ground to the treetops. After the sun vanishes below the horizon, masses of songbirds flutter up into the air and head north again. Whether by starlight or in pitch blackness beneath the clouds, the birds travel onward mile after mile, each wingbeat bringing them closer to the places where they know by instinct they must return.

For centuries, people around the world have watched and wondered about the seasonal movements of birds. Aristotle noted that some birds migrate from places nearby, but that others seem to come from the ends of the world. Despite this awareness, he wrote that redstarts of summertime turned into robins for the winter and that garden warblers changed into blackcaps. He also suggested that some songbirds, including dippers, larks, and swallows, survived the winter by hiding in holes in the ground. Similar beliefs persisted into the late 1700s, including the notion that swallows hibernated beneath water before emerging in spring.

As unlikely as that seems today, the truth is nearly as hard to believe. Cliff swallows from as far north as the Yukon in Canada fly to South America and back each year, a journey of more than 12,000 miles. Blackpoll warblers travel from Alaska to the Atlantic Ocean, then embark on a nonstop oceanic crossing of some 2,200 miles to South

America. During the thirty-six-hour flight over water, a blackpoll warbler travels so efficiently that researchers calculate it would log some 720,000 miles to the gallon if it burned gasoline instead of its own fat.[1] Across vast distances, songbirds navigate with remarkable precision, often returning to the same places each year on either end of their journeys. In addition to using stars and landmarks to find their way, songbirds orient themselves by sensing the earth's magnetic field and by checking their progress against patterns of polarized light that humans cannot see.

Songbirds are among the most familiar of birds in any landscape, inhabiting backyards, urban parks, and grasslands, as well as remote forests, deserts, and tundra. Of 10,000 bird species on earth, about 4,600 are songbirds. In North America, they include vireos, jays, crows, larks, swallows, thrushes, waxwings, chickadees, wrens, warblers, tanagers, grosbeaks, buntings, blackbirds, orioles, finches — more than 280 species in all. They belong to the order Passeriformes, or perching birds, but differ from their perching relatives, the flycatchers, in having complex voice boxes. With specialized muscles that help control sound, they produce varied and beautiful songs to attract mates and defend territories from rivals.

As familiar as songbirds may seem on their breeding grounds, surprisingly little is known about their lives during most of the year when migrants are traveling or wintering elsewhere. Since many songbirds migrate in the dark and travel immense distances, even basic information such as how many birds are aloft, how high they fly, and where they go has long been a mystery. For millennia, human knowledge of bird migration extended only as far as what people could see and hear. Only during the past fifty years have scientists gone beyond these limits with the aid of technology, exploring migration from the intricate details in a songbird's neural circuitry to the complex global picture of the birds' lives and travels.

This book follows songbirds through the four seasons as they migrate, mate, raise young, and winter in places thousands of miles away. It is a story of the journeys of birds as well as of the scientific breakthroughs that ornithologists have made in understanding the migrations that pass invisibly overhead. Each of the four parts concludes with a section offering tips about how and where to observe migratory songbirds using some of the same tools and techniques that scientists do. An appendix lists information about how to contribute bird sightings, whether from a backyard feeder or farther afield, to a growing base of scientific knowledge from birders around the continent.

In this age of technology, scientists have used weather radar to "see" for the first time how millions of birds cross over the Gulf of Mexico on a nonstop flight of six hundred miles. They have traveled by helicopter to live for months on narrow oil platforms in the Gulf, where they have witnessed migrations sweeping past at eye level in the glow of platform lights. On land, researchers have listened to the river of migration overhead, aided by computers that can harvest the sounds and record them as spectrograms that, like photographs, allow them to identify and quantify the passage of individual species indistinguishable on radar.

While some scientists have studied mass migration, others have followed the fates of individual birds. After attaching miniature radio transmitters to the backs of thrushes, researchers have followed the birds in darkness by car or by airplane for hundreds of miles. Using molecular techniques, they have measured how much energy a migrant expends during a night of flight. They have analyzed molecules in the blood and feathers of birds to trace where migrants have spent the winter. Researchers have also made fascinating discoveries about how songbirds navigate, through experiments with genetics, sensory perception, and neural responses in the brain.

On the geographical frontier, ornithologists have gathered thousands of records from citizen-science participants to understand continental movements of birds that winter north of Mexico. They have traveled to Latin America, where they have discovered striking similarities as well as remarkable differences in the way that birds live on their summering and wintering grounds. By studying Neotropical migrants in locales thousands of miles apart, they have revealed interconnections between the places migratory songbirds inhabit at all times of the year. With this perspective, they have shown, for example, what global warming on a mountaintop in Vermont could mean for thrushes that winter in the Dominican Republic and how events on a Caribbean island may influence the number of young a warbler can raise in the woods of New Hampshire.

They have also come up with some sobering calculations. For migratory songbirds, life is short and their journeys dangerous. Many songbirds live for less than two years, with more than 85 percent of annual mortality occurring during migration. Unfortunately, humans are making it increasingly difficult for birds to survive the trip. Hundreds of millions of songbirds each year die after crashing into windows and several million more are estimated to strike television and radio towers. Air pollution creates acid rain, which appears to be contributing to the decline of wood thrushes. Researchers worry that global warming may disrupt the precisely timed songbird migrations that coincide with the availability of food in spring and summer, or that changing temperatures may alter plant communities, making them unsuitable for some species of birds. Each year, as humans destroy more natural habitats, there are fewer places for birds to raise their young, to survive the winter, or to seek food and shelter throughout their journeys. The consequence is clear: the migrations of today are smaller than those in the past. Radar imagery suggests that migrations over the Gulf of Mexico may have diminished by as much as half since the 1960s and data on breeding birds show that the numbers of many songbird species have dwindled over the past several decades. This realization has mobilized international efforts to save the places on which migratory birds and other wildlife depend.

It may be that the remarkable amount of knowledge we have gained about songbirds has come just in time to help save them. More than ever before, we are able to appreciate the amazing phenomenon of migration, the resilience and vulnerability of birds, and the way that they connect us to places beyond our own borders. Understanding songbird migration is impossible just by trying to watch it; revealing its full scope and complexity has required ingenuity and imagination. The deepest appreciation of migration requires not only experiencing what we can of it, but also seeing it with our minds — and realizing that we have a role in protecting the far-flung landscapes that migratory birds need to survive.

SPRING

1
FLIGHT ACROSS THE GULF

A migrant song-bird I,
Out of the blue, between the sea and the sky,
Landward blown on bright, untiring wings;
Out of the South I fly . . .
Some irresistible impulse bears me on,
Through starry dusks and rosy mists of dawn,
And flames of noon and purple films of rain;
And the strain
Of mighty winds hurled roaring back and forth,
Between the caverns of the reeling earth,
Cannot bewilder me.
I know that I shall see
Just at the appointed time, the dogwood blow
And hear the willows rustle and the mill-stream flow.
—Maurice Thompson, from "Out of the South"

ON A FEBRUARY MORNING, the sun rises on the Gulf of Mexico. In the oak woodlands along the Louisiana coast, the songs of resident Carolina wrens mingle with sounds from the nearby surf. Boat-tailed grackles walk along the edge between the woods and the sand, and laughing gulls wheel effortlessly above the water. Beyond, the Gulf of Mexico stretches six hundred miles to another shore, where the waves wash up on the white sands of the Yucatán Peninsula. Here, Yucatán wrens live in the scrubby dunes, tropical mockingbirds perch in mangrove forests, and greater flamingos wade in the lagoons.

Soon, these year-round residents will be joined by migrant songbirds journeying between endpoints as far south as Argentina and as far north as the boreal forests of Canada. The journeys begin at nightfall in early spring, as bobolinks leave the marshes and grasslands of Argentina and scarlet and summer tanagers lift off from the montane evergreen forests of the Andes. Veeries depart from second-growth woodlands of the Brazilian Cerrado, dickcissels flutter up over the llanos of Venezuela, and Baltimore orioles fly over the open woodlands of Colombia. The migration swells each night, joined by migratory warblers, vireos, and thrushes that wintered with opal-crowned tanagers, blue-crowned motmots, white-plumed antbirds, and golden-headed manakins in the Amazon rainforest, or with crimson-fronted parakeets, bananaquits, and Montezuma oropendolas in the cacao plantations of Panama.

The migration gains momentum as millions of songbirds stream overhead — orioles, tanagers, thrushes, vireos, warblers, grosbeaks, buntings, bobolinks, and others — funneling northward through Central America and Mexico to the Yucatán Peninsula. There they come to the end of land and face two options: to follow the rim of land along the Gulf Coast that connects Mexico to the United States or to travel straight across the water with no land in sight. The songbirds head onward steadily, their paths obscured behind the curtain of night. By daylight, they arrive mysteriously on the northern Gulf Coast, in appearances that for centuries left naturalists wondering how they got there.

The earliest records of explorers to the New World hinted that songbirds must migrate over water, at least when close to land. In 1492, after sailing from Spain across the Atlantic for more than a month, Christopher Columbus noted many birds that stopped to rest on the ships or that passed overhead. On September 20, he recorded three pelicans, which he took as a sign of nearby land. Two or three landbirds also appeared near nightfall, sang aboard the ship, and disappeared before sunrise. On October 7, when flocks of birds passed overhead to the southwest, the restless crew changed course to follow the birds, hoping to find the shore. On October 9 they heard the calls of birds from above all night long, and on October 12, they finally sighted land.

More than three hundred years later, naturalist William Bullock recorded landbirds at sea that hinted at the passage of migrants above the Gulf of Mexico. In autumn 1824, numerous exhausted birds, mostly warblers and flycatchers, landed on his ship twenty-five miles off the coast of Campeche, Mexico, apparently on their way from the northern Gulf Coast to the Yucatán Peninsula. "The cabin was never without these pretty creatures, which entered the windows in pursuit of the flies," he wrote. "Many were of great beauty and variety, and some undescribed. Among those known were the purple heron, common snipe, pigmy sandpiper, the lesser spotted rail, American chatterer, orange and black warbler, and two kinds of swallows."[1]

During storms over the Gulf of Mexico, hundreds of birds sometimes materialized around ships. On April 2, 1881, Martin Abbott Frazar was on a small schooner about thirty miles south of the mouth of the Mississippi River when a sudden gale hit. Birds suddenly appeared from all directions, the winds forcing them down from the sky to just above the surface of the water. Frazar reported seeing large numbers of American redstarts, indigo buntings, prothonotary warblers, worm-eating warblers, Kentucky warblers, mourning warblers, hooded warblers, chestnut-sided warblers, and scarlet tanagers among the twenty-three bird species he recognized. Some landed on the deck but were swept overboard by the next wave. Others hit the sides of the ship and fell into the water. Frazar wrote, "It was sad indeed to see them struggling along by the side of the vessel in trying to pass ahead of her, for as soon as they were clear of the bow, they were invariably blown back into the water and drowned."[2]

In addition to records at sea, the whereabouts of birds along the coast in spring suggested that many of them must fly across the Gulf. In 1905, Wells W. Cooke noted in the journal of the American Ornithologists' Union, the *Auk*, that some birds, such as worm-eating warblers and chestnut-sided warblers, were virtually unknown along portions of the coastal routes that they would have taken if they had passed over land. Others, such as black-and-white warblers, arrived on the northern Gulf coast in spring before they appeared along the coast to the south, evidence that some of them had taken a quicker route over water.

Cooke's findings went largely unchallenged until 1945, when George G. Williams insisted in the *Auk* that Cooke's records were unconvincing. There was no direct evidence that birds were crossing the Gulf of Mexico in appreciable numbers, Williams argued. Birds seen from ships during storms could have been blown off course into the Gulf. Indeed, although most ornithologists believed that millions of migrants crossed the Gulf each spring and fall, no one had ever witnessed trans-Gulf migrations on the scale to be expected if such migrations really occurred. George Lowery, an ornithologist from Louisiana State University, admitted that trans-Gulf migration was fraught with enigmas and strained one's credulity, but he considered those to be poor excuses for ruling out the possibility.

On April 24, 1945, Lowery boarded the S.S. *Bertha Brøvig* in New Orleans, bound for the Yucatán Peninsula in search of migration over the Gulf. That night, he set up his telescope on the deck and aimed the lens at the moon. As Lowery peered through the telescope's eyepiece, he hoped to see the silhouettes of birds crossing in front of the moon, but no matter how he steadied himself against the rocking of the boat, the moon kept bobbing in and out of view. With the telescope useless at his side, Lowery could only have gazed up in the darkness, knowing that thousands of birds could be passing between him and the moon, but that there was no way to see them.

Undaunted, Lowery was back on deck at first light, scanning the sky and meticulously recording the ship's coordinates each time he saw a bird. "April 30, 1945. 4:45 A.M.," Lowery logged in his first entry. "One mile off South Pass. Two yellow warblers (_Dendroica petechia_) and a thrush (_Hylocichla sp._). Seen approaching from the south; they passed alongside of the ship and continued northward in the direction of land."[3] Half an hour passed without any other bird sightings, but at five fifteen, Lowery recorded a small bird, possibly a female indigo bunting, circling the ship three miles offshore. As the _Bertha Brøvig_ progressed southward, he saw other birds at intervals, including a female dickcissel, an eastern kingbird that perched on a rope, and an immature male orchard oriole that landed in some hay on the deck. On May 1, Lowery recorded additional birds, including three barn swallows, 248 miles from Louisiana and 282 miles from the Yucatán coast. Upon completing the round-trip voyage, Lowery tallied sixty-one landbirds of twenty-one species over the Gulf.

Admittedly, the numbers were small—he had seen fewer than six birds per day on the return trip—but Lowery believed that was to be expected. Birds migrating over land flew at altitudes too high to be seen, except when bad weather brought them down. He assumed that the same must be true for birds migrating over water. He wrote that observations of birds in the Gulf during fair weather could be nothing short of astonishing to anyone familiar with migration over land.

However, it was his observations on the Yucatán Peninsula that Lowery believed would convince skeptics of the magnitude of trans-Gulf migration. After the ship docked at Progreso, Lowery set up his telescope on land. This time he saw and counted the silhouettes of birds as they flew past, backlit by the glow of the moon. With these numbers, he calculated how many birds must be aloft, based on the area viewed through the telescope's field—a slender cone about thirty feet in diameter at an elevation of two thousand feet.

During his first watch between two forty-five A.M. and four A.M. on May 5, Lowery counted the silhouettes of twelve birds as they passed in front of the moon. Based on this number, he calculated that 3,710 birds were crossing a one-mile line in a single hour, heading north over the Gulf. The next night, he watched through the telescope again between three twenty A.M. and four twenty A.M. and counted eight silhouettes, representing some 1,960 birds crossing a one-mile stretch in the course of an hour. These numbers convinced him that migrants were passing over the coast of the Yucatán by the hundreds of thousands. In a 1946 report in the *Auk*, Lowery concluded that the flight densities at Progreso indicated that birds seen over the Gulf were merely the visible fraction of an immense flight passing overhead.

Williams took a different view. In a paper published in the *Auk* the following year, he accused Lowery of jumping from minute observations to sweeping conclusions. He pointed out that calculations from Lowery's formulas indicated that 21 million birds passed north from the peninsula on the first day of his observations—on the basis of twelve birds seen. He wrote, "Even if we ignore the many other questions involved in the observations, Lowery's method of estimating millions of birds on the basis of the minute amount of birds he actually saw is so fabulously hypothetical that I cannot see that it has any value at all."[4] Williams also dismissed Lowery's sightings of birds at sea. He argued that the kingbird and barn swallows were confused by smoke from fires that obscured the coast and had strayed from their overland route. The records proved, he said, that birds over the Gulf in spring amounted to a few mere droplets compared with the floods of birds that must go around the Gulf.

A few mere droplets — no one could deny that the direct evidence for trans-Gulf migration was less than satisfying. A handful of flecks crossing the moon, the occasional voices of birds passing overhead in the dark, or even hundreds of birds taking refuge on ships during a storm, were the limits of what human perception allowed anyone to see. In the years that followed, Lowery and his colleague Robert Newman continued to gather moon-watching data and to record birds that were visible through telescopes against a backdrop of clouds. But it took a new technology — radar — and an astute teenager, Sidney Gauthreaux, to reveal the full sweep of millions of birds flying in from the Gulf.

AS A YOUNGSTER growing up in New Orleans, Gauthreaux had witnessed some of the most spectacular fallouts of migratory birds on record. On May 5, 1957, a powerful cold front from the north brought thousands of birds down from the sky. "There was a sudden appearance of incredibly colorful birds, like painted buntings and indigo buntings, Baltimore orioles, and lots of tanagers," Gauthreaux said. "I was seeing warblers by the hundreds — maybe 500 ovenbirds, 350 bay-breasted warblers . . . It was all over the city of New Orleans, no matter where you traveled, what woodland you checked. Even people in church couldn't help but notice all the colorful birds coming down outside the windows."[5]

Witnessing these events solidified Gauthreaux's desire to become an ornithologist and study bird migration. When he was a teenager, he began using Lowery's moon-watching technique to record migrants at night. He also discovered that if he pointed his telescope straight up against a layer of high cirrus clouds, he could see what looked like pepper flakes — tiny silhouettes of migrants coming in off the Gulf by daylight — a technique that he later learned was independently discovered by Newman at about the same time.

In 1957, when Gauthreaux was seventeen years old, Louisiana received some of the first weather radar stations in the United States, deployed as warning systems for hurricanes. Called WSR-57, for Weather Surveillance Radar 1957, the new radar images were broadcast on Louisiana television. The newscaster explained how meteorologists could detect the raindrops of approaching storms, represented by white dots on the radar screen. When Gauthreaux saw the images, he noticed, among the fuzzy dots that represented rain, something that he thought must be the movements of birds.

"I saw little targets going back and forth on the lakefront of New Orleans that I was absolutely certain were things like gulls flying back and forth," Gauthreaux said. "After all, a weather radar functions by picking up minute droplets of water in the atmosphere. So it made perfect sense that radar designed to pick up small droplets of water would likely pick up migrating birds."

Indeed, radar picked up birds so well that it had caused considerable confusion for the military before radar operators learned to identify stray signals as birds. In a 1946 issue of the *Auk,* I. O. Buss described how, throughout the war, sudden and mysterious radar signals rushed combat men to battle stations, sent fighter planes on goose chases, and set off at least one invasion alarm.

As a teenager, Gauthreaux didn't know that the military's mysterious radar signals, dubbed "angels," had turned out to be birds. Hoping to learn whether radar could pick up migrant birds coming in off the Gulf, he went to the local weather station to ask if the meteorologists would let him take a closer look. He was fascinated by what he saw on his first visit and returned repeatedly to pore over more radar images. On nights when he could see birds flying in front of the moon through his telescope, he also saw dots on the radar screen, the echoes of the radar microwaves that were hitting the birds and bouncing back to the radar receiver to be detected at the station. "I realized that here's the magic eye that can see everything in the sky—and can even see through the clouds," Gauthreaux said. In spring, when he knew that migrants were coming in from over the Gulf, the birds were so numerous that they sometimes whited out the display on the radar screen. "It was obvious to me that it was picking up trans-Gulf migration."

It wasn't as obvious to everyone though, not even to Lowery, who should have been elated about the potential for radar to document trans-Gulf migrations. Although Gauthreaux had met Lowery and Newman during his teen years on bird watching excursions, he first approached them about using radar for studying trans-Gulf migration when he applied to study with Lowery in graduate school. "Initially Lowery and Newman didn't think that radar was picking up birds," Gauthreaux said. "They were a little skeptical about the new technology."

Nevertheless, Lowery accepted him as a student, and Gauthreaux embarked on a master's thesis that would eventually convince Lowery and Newman that radar could detect trans-Gulf migration. By counting birds using telescopic techniques and comparing the numbers with data from radar images captured at the same time, Gauthreaux showed that the radar data corroborated calculations based on moon-watching, across a greatly expanded panorama. During the next several years, he continued studies of migration using radar for his doctoral dissertation with Lowery as his advisor.

By watching the radar screen, Gauthreaux could detect birds from fifty to seventy-five nautical miles away as they approached the coast in numbers as high as fifty thousand across a one-mile stretch. Typically the radar showed birds at elevations of 1,900 feet to 8,200 feet, or sometimes as high as 12,000 to 15,000 feet, more than two miles above the sea. Gauthreaux presumed that the altitude depended on where the winds were favorable. His calculations showed that, in the absence of tailwinds, birds leaving the southern Gulf coast would run out of fuel before reaching the other side. On the other hand, birds that rode the wind over the Gulf would arrive with energy to spare.

Most of the time, the massive arrivals of birds went unnoticed along the beaches and coastal woodlands, the first land that the birds encountered after their nonstop flight of fifteen hours or more. Radar showed specks that were birds, continuing without pause beyond the first twenty to thirty nautical miles of marsh, prairie, and scattered woodlands, pressing on toward inland forests. Severe storms, however, caused the specks to disappear from the radar screens. When that happened, Gauthreaux would step outside to see flocks of migrants plummeting out of the sky and landing in trees in spectacular concentrations. Once the storms passed, usually less than three hours later, radar showed the birds moving inland again.

The birds' reliance on wind explained many of the seasonal and daily patterns that Gauthreaux recorded using radar. He found that migration occurred almost every day from April 8 to May 15. Winds over the Gulf in March generally flow from the east, but in April and May, winds become progressively favorable and fewer cold fronts pass through. With southerly winds of ten to fifteen miles per hour, large numbers of migrants would begin to arrive around ten A.M., peak in midafternoon, and taper off until nightfall. In headwinds, peak numbers of migrants did not arrive until five in the evening and many migrants continued arriving after nightfall.

Thus, on most nights, migrants could make it safely across the Gulf with the help of following winds, though they risked encountering cold fronts that would sap their energy before they reached land. Gauthreaux recalled standing on an oil rig fifty-five miles from land and seeing some birds that were so tired they couldn't make it to the rig and plunged into the water. Others died after landing on the rig. "The birds' breast bones were sharp like a knife blade," Gauthreaux said. "They had basically burned up all their fat and most of their muscle mass to stay aloft. One would have reasoned that with that kind of mortality, trans-Gulf migration would have been selected against. Not so. The benefits of flying over the Gulf, as opposed to around it, must be incredible for this behavior to persist."

Traveling across the Gulf is faster than traveling around it, and that probably explains why so many migrants risk the journey. Most migratory songbirds fly at night, after the earth cools and the atmosphere becomes less turbulent — an important energy-saving measure. By flying in darkness, birds can also avoid diurnal aerial predators such as hawks. With tailwinds, birds crossing the Gulf can reach the northern coast in a single night, whereas birds going the longer route around the Gulf would take five or six nights to travel to the same point. Cutting down on the amount of travel saves energy, as well as the amount of time exposed to predators along the way. Birds arriving early on their breeding grounds gain other payoffs too: they can lay claim to the best territories and have a better chance of raising more young than birds that arrive late.

With new radar technologies, the potential for understanding migration continues to expand. Gauthreaux is now a professor of biology at Clemson University, where he heads the Clemson University Radar Ornithology Laboratory. In 1992, Gauthreaux and colleague Carroll Belser tried out a new Doppler weather surveillance radar (WSR-88D) south of Houston, Texas. The new radar was more sensitive than the older WSR-57 and displayed new information such as the direction and speed of bird movements, in color. Within a few years, ten WSR-88D stations were located from Brownsville, Texas, to Key West, Florida, allowing Gauthreaux and Belser to study migration across the entire northern Gulf coast.

One sobering result of recent radar work is that the

volume of trans-Gulf migration appears to have diminished during the two decades after Gauthreaux first began collecting radar data. Gauthreaux found that the percentage of spring days with arriving trans-Gulf flights during 1987 through 1989 was only about half of that during 1965 through 1967. His subsequent analyses have shown that the density of migrants has also diminished. Which birds have been affected, and why? One of the limitations of radar is that it cannot distinguish one species from another. Observers on the ground have shown that many trans-Gulf flights landing along the coast are composed of shorebirds. "If you start checking shorebird areas during some fallouts, there are shorebirds coming in all over the place," Gauthreaux said. "But folks looking at the woodlands for songbirds are seeing hardly anything." This means that songbird numbers may have diminished even more than radar images have been able to suggest. "Even right now we're still discovering things about trans-Gulf migration that will require a good deal more work," Gauthreaux said.

SOME OF THAT WORK has involved returning to the Gulf with binoculars and telescopes. In 1998, a group of ornithologists from Louisiana State University initiated the Migration over the Gulf Project in cooperation with oil companies BP Amoco, ExxonMobil, Newfield Exploration, Phillips, Shell, and Texaco to monitor migration from oil platforms. It was the first time anyone had used oil platforms to study trans-Gulf migration at length and to document how the artificial archipelago of some four thousand platforms might be affecting the migration of birds.

When John Arvin, a lifelong birder and professional

bird watching tour guide, learned about the project, he jumped at the chance to witness migration at closer range than he ever had before. "What bird watchers see when they rush to the Gulf coastal groves and woodlands that serve as well-known 'traps' for migrants is a scrap of the results of a migration, not the migration itself," Arvin wrote in *Texas Birds*. "My lifelong fascination with the migration of small landbirds stems largely from the fact that it is almost completely invisible, and therefore somewhat mysterious and poorly understood."[6]

Arvin was one of five observers assigned to different oil platforms during the spring of 1998. Before being transported by helicopter to the rigs, the five Birdmen, as they were called, had to pass a three-day intensive training session that included horrific movies of platform blowouts at sea and crashes of helicopters transporting personnel to the platforms. They spent half a day in a frigid swimming pool, where they were blindfolded and spun around in a helicopter simulator underwater. They were required to escape from it in preparation for an actual crash. "It made you start to consider, " 'Why did I think this was a good idea in the first place?' " Arvin recalled with amusement.[7]

On March 12, the helicopter dropped Arvin off without incident on an oil platform eighty miles due south of Lafayette, Louisiana. Home was now a vertical steel structure surrounded by water and sky. The platform was a series of decks with oil production equipment and miles of piping that went around the 115-foot tower. About eighty feet above the water's surface stood a small building that Arvin shared with the two crews who kept the equipment running to extract oil from the well below. The housing included bunk rooms, a galley for cooking, offices, and a recreation room — not that Arvin would need exercise after walking up and down some five thousand stairs each day to census birds along the platform.

According to protocol, Arvin and the observers on the other platforms spent seven hours each day counting birds on the platform and in the air, beginning at five A.M. and ending at nine P.M. They wore the garb required of everyone on the platforms — life vest, hard hat, steel-toed boots, goggles, and earplugs. Accompanied by the loud whines and clanging of motors and generators in the background, they covered a one-hour census route along the entire platform five times each day. For two sessions each day they took up a fixed position and scanned for birds flying past. They completed the routine even on days when they saw no birds at all.

On the platforms, as elsewhere on ships or land, fair days brought few hints of migrations that might be passing overhead. Four weeks passed before Arvin was able to hear the calls of any birds above the constant noise of the platform's machinery. "It was the first time in my life that I'd gone for a month without hearing a bird," Arvin said. "I realized then how it was real deprivation not being able to hear them."

The isolation of life on the platforms was also difficult, but Arvin exchanged notes by phone with the Birdmen stationed on other platforms. For the most part, in good weather, birds simply weren't landing on the platforms. Those that did come down were in such poor condition that they often ended up dying there.

In strong winds or rain, greater numbers of birds landed.

Bob Russell, on another platform, noticed that birds used parts of the platform that resembled their natural habitats. Wrens perched in piles of loosely strewn wire that resembled brush piles, but canopy birds rested in the piping above. Birds that fed in open areas spent their time on the decks as if in an open field. Catbirds took shelter in low, complicated production equipment that suggested a thorny scrubby area, while marsh wrens preferred the grating on walkways near the water that substituted for matted-down marsh grass and reeds. Usually the birds stayed only until the weather let up and they could continue their journey.

Arvin's first real glimpse of bird migration occurred on March 31, when sudden winds brought masses of birds down to eye level. He had just finished lunch when he walked outside and saw, without binoculars, a dark line approaching from across the Gulf. "Right in front of this wall of weather the air just filled with birds instantly," Arvin said. "I raced off to get my binoculars."

"The whole rest of the day the air was full of birds," Arvin recalled. "There were a lot of herons and ducks beating their way directly into a north wind. Their progress toward us was extremely slow and sometimes they would get caught in the turbulence of air flowing around the platform and get hurled back a mile or so, then have to do it all over again. Under really windy conditions, most of them would get down real low where the friction from the water decreased the wind speed a little. You'd think it would be dangerous down that close to the water. I imagine that occasionally some of them were lost when they got swamped by a wave."

In the high winds, a flock of at least fifty brown-headed

cowbirds stopped to wait out the gusts, along with a yellow-headed blackbird. Because these birds winter to the north or to the west of the Gulf, Arvin was surprised to find them on the platform eighty miles south of the nearest land. These short-distance migrants probably had been aloft and had been blown out to sea as they were migrating over land — an observation that trans-Gulf skeptic George Williams would have appreciated.

By late April, most of the migrants were songbirds. In good weather, Arvin would go up to the helicopter deck at the top of the platform just after sunset. Lying on his back and looking through binoculars, he could see streams of songbirds passing overhead at the limit of vision, illuminated from below by the sun's rays beneath the horizon.

On April 28, weather brought the migrants down once more, when fierce winds began to whip around the platform at nightfall. Although lightning flashed in the distance, Arvin could see only about fifty feet into the darkness, in the space illuminated by the glare from the platform's lights. He was astonished to see some two hundred warbler-sized birds flying into the wind about eighty feet above the water. Though they made small movements from side to side and up and down, they appeared to be flying in place — except when the wind hurled them back into the darkness. Arvin had to struggle just to hold his binoculars steady in the wind, and the harsh light washed out the birds' colors and markings. Some, though, flew close enough for Arvin to recognize as they fluttered in place. Most were bay-breasted warblers. He also identified ovenbirds, scarlet tanagers, and rose-breasted grosbeaks. At eleven P.M., the torrents of rain made it impossible to stay out any longer. By five A.M., the start of the day's first census, the wind had died down and the skies had cleared. A few dead male bay-breasted warblers, victims of collision with the platform, were among the only evidence of the previous night's drama.

On the platform to Arvin's west, near the Louisiana-Texas border, the storm had caused an even more spectacular fallout. After nightfall, more than ten thousand birds began circling the platform, apparently attracted to its lights. Flying counterclockwise, they continued circulating until the glow of early morning drew them out of the column of artificial light. During the five A.M. census, observer Jon King reported that the south subcellar deck was carpeted in warblers, at least six hundred bay-breasted warblers among them. He counted more than two thousand warblers on the platform in all.

On that day, April 29, the daylight hours passed uneventfully on John Arvin's platform, until the sunset had faded to a glow in the west. At about eight P.M., where the platform cast its light, Arvin noticed one bird after another flying into the illuminated space, then disappearing into the darkness. They were heading north into a light headwind — most of them thrushes. They flew so close that Arvin could hear the constant calls of gray-cheeked thrushes as they flew by in staggering numbers. Scarlet tanagers, summer tanagers, rose-breasted grosbeaks, blue grosbeaks, gray catbirds, Baltimore orioles, yellow-billed cuckoos, red-eyed vireos, and a variety of warblers — yellow, blackpoll, prothonotary, hooded, bay-breasted, and American redstarts — flew past steadily by the thousands, from a few feet above the water to fifteen feet above Arvin's head where he stood at the top of the deck.

"They streamed by just feet away but made no attempts to approach or land," Arvin wrote. "I could have caught dozens, perhaps hundreds, with a butterfly net."[8] A few hours later, Arvin moved to a well-lighted deck on the north side of the platform, "Here, it was possible to stand and have the flow of birds barely part enough to avoid colliding with my body and then close ranks again once past me," he wrote. "The effect was exactly like standing on a rock in the middle of a swiftly flowing river."[9] The river of birds continued to flow hour after hour, as he watched.

Arvin estimated that hundreds of thousands of birds, perhaps millions, passed his platform that night. By five A.M., the stream had slowed to the last stragglers. About fifty birds had become confused after flying beneath one of the decks and were still trying to reach the sky by fluttering against the steel mesh ceiling. Others were apparently sleeping, their tails protruding from the beams. Several blackpoll warblers were hopping on the decks and catching insects attracted to the platform's lights. By six, the procession of birds had stopped or ascended too high to be seen, and the birds on the platform had dispersed.

One of the men on the crew said he had seen nothing like it in all the fifteen years that he had worked on the platform. A rare combination of weather patterns must have converged on a narrow portion of the Gulf that night, causing the birds to fly near the surface of the water in apparently benign conditions. Observers to Arvin's east hadn't noticed any extraordinary flights.

"This has been the ornithological spectacle of my entire life," Arvin wrote in an e-mail message to his colleagues the next day. "I'm a little rocky from no sleep but I have never been remotely struck by any other ornithological event like I have been by this. I suspect that very few people on this planet have seen what I have seen in the last 12 hours. It is a spell-binding feeling I will carry to my grave."[10]

The project continued each spring and fall through 2000. No one saw another spectacle like the ones that Arvin and King had witnessed, though there were numerous other revelations. At the completion of the project, the stack of data sheets stood more than six feet high. These records included 120,000 case histories of birds, including observations of when each bird arrived on the platform, how long it stayed, whether it was foraging, and its condition upon leaving.

One finding was that the platforms were not causing mass mortalities of birds, as some had feared before the study was conducted. Some birds died from collisions with platforms when the wind hurled them too close. Others simply died of exhaustion, having burned up all of their fat and then their muscle tissue. But these numbers were small—about 15 dead birds per season on a platform, or rarely as many as 150. Arvin picked up 34 dead birds from the platform during the spring of 1998.

Additionally, some birds probably benefit from platforms. "We've seen dozens or hundreds of birds resting on platforms that otherwise undoubtedly would have died in the Gulf," Russell said. In spring, birds passing the northern platforms, where the project's observers were stationed, were near the end of their journeys. In some cases, the birds might never have completed the trip if they had not been able to use the platforms to rest and refuel. In fall, birds sometimes stop on platforms during "reverse migrations" caused by bad weather. On October 2, 1998, a large migratory flight headed south over the Gulf with favorable tailwinds but then encountered headwinds partway across. Apparently, after flying into the headwinds for hours, they aborted the flight and rode the air currents back to the north, arriving exhausted on the platforms on the afternoon of October 3.

Most migrants, though, bypass the platforms altogether, and that, perhaps, is the most striking message of all. The view from oil platforms, Gauthreaux said, drives home that "there is no question that trans-Gulf seems to be a very normal operation. Tremendous numbers of birds fly overhead as successful migrants and make it into the United States."

However awesome the recent glimpses of trans-Gulf migration have been, this means that scientists still know little about migration as a routine journey for millions of birds, some of which have crossed the Gulf of Mexico numerous times in their lives. Limited to what we can see from just above the surface of the water, we know these mass migrations fleetingly, and only when the birds have descended because of freakish weather conditions. And despite the startling radar images that prove the Gulf is no obstacle to millions of birds, the images are, after all, just dots, indistinguishable from one another.

For now, the scientific frontier seems to be aloft in a layer of the atmosphere that, if not for radar, would be virtually unknown to biologists. Which species are among the stream of birds from night to night as spring advances? Are they stratified in the sky according to species, sex, or age? Why do some species use nocturnal flight calls while they migrate? "We get birds coming in off the Gulf at 12,000 feet and 15,000 feet above the ground," said Gauthreaux. "We still don't know the answers to questions even as fundamental as the species of birds involved. I would absolutely love, somehow, some way, to get up there and identify the birds. That's something I want to stress to students of bird migration. Don't get the impression that we've found all the answers. We're just now beginning to scratch the surface."

2
MAKING LANDFALL

Far in the south, where vast Maragnon flows,
And boundless forests unknown wilds inclose;
Vine-tangled shores, and suffocating woods,
Parched up with heat or drowned with pouring floods,
Where each extreme alternately prevails,
And Nature sad their ravages bewails;
Lo! high in air, above those trackless wastes,
With Spring's return the king bird hither hastes;
Coasts the famed Gulf, and from his height explores
Its thousand streams, its long-indented shores,
Its plains immense, wide op'ning on the day,
Its lakes and isles, where feathered millions play.
All tempt not him; till, gazing from on high,
COLUMBIA's regions wide below him lie;
There end his wanderings and his wish to roam,
There lie his native woods, his field, his home;
Down, circling, he descends, from azure heights,
And on a full-blown sassafras alights.
  —Alexander Wilson, from "King Bird"

I N JOHNSONS BAYOU, Louisiana, where surf from the Gulf of Mexico washes up on the beach, sanderlings and piping plovers run along the sand, probing with their bills for invertebrates. A flock of laughing gulls rests and preens where the high tide has deposited piles of driftwood and drying kelp. Tree swallows dip and skim through the air, traveling back and forth over water, sand, and the abrupt tangle of green vines and brush at the edge of the dunes. Just beyond, clinging to the ancient ridgeline, gnarled oaks and honey locust trees spread their branches over mulberry bushes and honeysuckle vines. A water moccasin darts across a dirt path and into a sunlit puddle then raises its head, opening its white mouth to the sky. Above, six white ibis soar effortlessly over the treetops, crossing over the narrow patch of woods to the marsh on the other side in a matter of seconds.

On spring days when thousands of songbirds fly in from the Gulf, this is the first land they cross after fifteen hours or more without rest or food—a ribbon of sand, a belt of trees, and an open marsh beyond, as far as they can see. In good weather, when winds blow from the south, virtually all of the songbirds keep flying, passing over the inhospitable beach, the five-hundred-foot-wide woods, and at least thirty miles of marsh, to the bottomland hardwood forests beyond. On those days, few migrants stop in the coastal woods, though the air is sweet with the smell of blooming flowers, the mulberry vines are laden with purple fruit, and caterpillars crawl among the canopy leaves.

When the winds shift to the north, forcing exhausted migrants to fight against headwinds, or when sudden rains hit the coastline, the cheniers, or "places of oak," come alive with songbirds. During these fallouts, thousands of colorful birds funnel into the cheniers. Depending on which birds are aloft in greatest numbers when the weather hits, the sky may rain catbirds, send down flurries of magnolia warblers, or sprinkle with indigo buntings.

Since 1985, biologist Frank Moore and his research team from the University of Mississippi have been studying migrants that make landfall in the cheniers after their six-hundred-mile flight across the Gulf. By observing the birds when they land and using mist nets to capture them and take measurements, the researchers are learning which birds stop and why, and how the migrants cope in an unfamiliar place where they must seek shelter, avoid predators, and find enough food to fuel the next leg of their journey.

On May 1, 2004, one of Moore's graduate students, Zoltán Németh, was working in the chenier with a crew of field assistants when the rain and wind forced them to close the nets. Soaking wet, they headed back to the beach house to wait out the storm. That afternoon, as Németh was reading a book in the living room, he heard a sudden "thunk" from the porch window facing the beach. He looked up just in time to glimpse a tiny greenish bird bouncing off the window screen. As the bird flew unsteadily around the side of the house and out of sight, Németh sprang off the couch, grabbed his binoculars, and ran out the front door. His field assistant Ashley Sutton was standing outside smoking a cigarette as the wind gusted past the house and across the sandy beach beyond. "Did you see what hit the window?" Németh asked. "Tennessee warbler," Sutton said.[1]

Then Németh noticed the flitting movements of warblers

in the trees. He saw at least one Tennessee warbler and counted four American redstarts, their black and orange plumage flashing among oak leaves. A dash of blue, green, and red caught his eye — a painted bunting. After some eighteen hours aloft, these birds had barely made it across the Gulf of Mexico and a few hundred yards of sand before landing, exhausted, by the beach house.

Németh, Sutton, and field assistant Amy Tegeler ran down to the beach to look for migrants coming over the water. As they stood on the sand, looking out at the blue expanse, they saw a dark speck flitting toward them. It came in low to the water, fighting the headwinds, and passed them so closely that they didn't even need to lift their binoculars to see it was a black-and-white warbler. Moments later, another bird flew toward them just above their heads — an American redstart. A magnolia warbler beat its way past, then a hooded warbler, a few brilliantly blue indigo buntings, and several thrushes. A few of them landed on the porch of a nearby beach house; the others maneuvered between the houses and landed in the trees beyond. Then Sutton pointed out over the water and counted six grayish birds with narrow wings approaching the beach. "They're nighthawks!" he exclaimed. The nighthawks flew past rapidly and out of sight.

Németh, Sutton, and Tegeler raced back to the beach house and piled into the car, along with field assistant Jimmy Gore. They drove along the two-lane road paralleling the beach for about four miles to the town of Johnsons Bayou. Németh pulled over at a gas processing plant surrounded by barbed wire. He punched in an access code at the gate then drove past a maze of gray buildings, smokestacks, and tubes, toward the beach to the edge of the woods, but by then it was raining again. They would have to wait until the next day to find out what else the weather had brought down.

When Németh and his crew arrived at the chenier the next morning, a chilly north wind was still sweeping through the trees, but the woods were seething with birds—all of them looking for food. In a towering honey locust tree, a bay-breasted warbler hopped along a branch, cocking its head from side to side in search of insects. A Blackburnian warbler, in blazing orange and black plumage, hovered at the tip of a branch to glean insects from the leaves as a black-and-white warbler crept along a branch upside down, probing the bark with its bill. Flocks of gray catbirds leapt up from an overgrown mulberry bush, then fluttered back down to pull purple mulberries from the vines. On the ground, earth-brown thrushes were turning over the leaf litter in search of invertebrates, and ovenbirds were hopping up to pick insects from overhanging shrubs. The live oak trees seemed filled with birds of all colors—black-throated green warblers, rosy-red summer tanagers, indigo buntings, common yellowthroats, rose-breasted grosbeaks. And everywhere, it seemed, yellow and black magnolia warblers were hovering, flitting, and gleaning insects from the undersides of leaves. Above the rush of the wind, the melodies of hooded warblers, wood thrushes, and painted buntings mixed with the calls of magnolia warblers, American redstarts, and gray catbirds.

Fanning out along the trails in the twenty-acre patch of woods, they quickly unfurled the twenty-four black mist nets that had been closed since the day before. Birds began hitting the nets before they were even fully open — an ovenbird, a Swainson's thrush, a northern waterthrush, a rose-breasted grosbeak, a magnolia warbler. Tegeler, Németh, and David Rios worked as quickly as they could to untangle the birds. They placed each bird into one of eight compartments in a box the size of a small suitcase. When all the compartments were filled, the box resembled an apartment building with its bird occupants looking out the mesh windows. Rios walked quickly along the footpath and delivered the box to the banding area, a picnic table beneath the wide canopy of an open tent.

As Erica Henry waited with pencil poised to record data, Sutton reached into a compartment and pulled out a chestnut-sided warbler. Twenty-four hours earlier, the tiny bird might have been snatching insects from a blooming inga tree in Mexico; now after having flown across the Gulf, it looked remarkably unruffled, from its yellow cap to its pure-white breast bordered by bold chestnut stripes.

The bird gazed up with alert black eyes as Sutton used banding pliers to close a shiny new numbered aluminum band around the bird's leg. "Male, after-second-year," Sutton said, announcing the bird's sex and age. Then he blew on the breast and belly feathers to look for fat beneath the translucent skin. "Fat two," he said, rating the bird's fat score on a scale of zero to six. "Not enough to make it across the Gulf but he doesn't have to worry about that since he already made it." Quickly, Sutton scored the bird for the amount of breast muscle visible on either side of the breast bone, an indicator of the bird's condition. He measured the wing and tarsus, then placed the bird head-down in a canister on an electronic scale. "He's 9.5 grams," Sutton said.[2] Then he tilted the canister, gently removed the bird, and set it free. The entire process took less than one minute.

In nearly twenty years, Frank Moore's research group has measured and weighed thousands of birds in Johnsons Bayou and other sites along the Gulf Coast. About 40 percent of wood-warblers in the cheniers score zero for fat. Some birds, after having depleted their fat reserves en route from Mexico, have also burned up their muscle tissue as a last-ditch source of energy to get them across the Gulf. About one in every four thrushes weighs less than a typical fat-free thrush.

For birds out of fuel, the cheniers are a life-saving place. Among the oak and honey locust trees and the dense tangles of Japanese honeysuckle, wild grape, and mulberry, the birds find shelter, nectar, fruit, and insects. Measurements of birds recaptured in the cheniers from one day to another show that they can regain fat at about 10 percent of their lean body mass per day—the equivalent of 15 pounds a day for a 150-pound person.

On days when the weather forces them down, more of the birds are in better condition than on days that are favorable for migration. That's because when south winds blow steadily in from over the Gulf, all of the songbirds, except those in the worst condition, fly over the strip of chenier and travel at least another thirty to forty miles to reach the bottomland hardwood forests, where food and shelter are more plentiful. Like 35 percent of all wood-warblers captured in the cheniers, Sutton's chestnut-sided warbler had enough fat that it probably would have kept migrating if not for the twenty-mile-per-hour headwinds.

All morning, the banders pulled out a succession of

stunningly beautiful birds from the boxes, one after another — scarlet tanager, gray-cheeked thrush, yellow-billed cuckoo, hooded warbler, red-eyed vireo, magnolia warbler, Kentucky warbler. So many birds flew into the nets that by noon all of the holding boxes and the back-up supply of cloth bags were filled with birds, so the banders had to close the nets. Two hours later, they reopened the nets and kept banding birds until nightfall, when Sutton, wearing a headlamp, released the last bird of the day — a female magnolia warbler. In ten hours, they had captured 438 birds, shattering the previous daily record — 345 — in eleven years of mist netting at Johnsons Bayou.

This extraordinary fallout had occurred near the peak of spring migration on the Gulf Coast — late April and early May, a time when trees are leafing out after the winter and winds across the Gulf are increasingly favorable for migration. On April 30, millions of songbirds must have been surging northward from Mexico, riding winds from the south. It wasn't until they were partway across that they encountered a powerful cold front blasting toward them.

On the morning of May 1, when Németh's team closed the nets in Johnsons Bayou in a downpour, some of those migrants must have been fighting their way into headwinds, mustering their last energy to get across the Gulf. Later that day, crossing over the last stretch of water, they would have seen the thin strip of sandy beach and the belt of woodlands clinging to the beach ridge. Some, flying low to the water, would barely make it, too exhausted to maneuver around the beach houses that stood between them and the trees. Others, seeing nothing but marshes beyond the corridor of greenery, would descend by the thousands into the chenier.

Although the Gulf Coast is renowned for its fallouts, the number of migrants in the cheniers is as variable from day to day as the spring weather. On rare days during spring migration, all the nets at Johnsons Bayou stand empty from dawn to dusk. The kinds of birds in the cheniers also vary from day to day depending on which species happen to be migrating when the weather brings them down. Of the 438 birds captured on the record-breaking day, 101 were magnolia warblers. (Previously, the highest number of magnolia warblers captured for an *entire seven-week season* was 53.) The fallout occurred in early May, a time when magnolia warblers migrate over the Gulf Coast in large numbers. The same weather conditions in early spring might have ushered in hundreds of hooded warblers, Kentucky warblers, or common yellowthroats instead. Depending on the timing of winds and rain, banders may see multitudes of a particular species in one year, and hardly any in another. For example, in 1993, a research team led by Wylie Barrow from the USGS Wetland Center captured 489 yellow-rumped warblers all spring; in 2004 they only captured 1. During the spring of 2003, they captured 201 hooded warblers, but only 7 in 1994.

The winds influence not only whether the birds will land, but also where they end up when they do. Migratory songbirds have remarkable homing abilities, and recaptures of birds along their migratory routes over land have shown that they often follow the same pathways from one year to another. In eleven years at Johnsons Bayou, however, Moore's group has never caught the same songbird twice in migration during different years. This means that nearly all of the birds are landing in unfamiliar terrain—at a time when they are particularly vulnerable because of hunger and exhaustion.

When they first arrive in the cheniers, the birds sometimes fly quickly in the treetops, as if assessing their surroundings. With fervor, they begin looking for food, sometimes even abandoning their usual foraging methods as if to try anything that works. Diane Loria and Frank Moore compared the feeding strategies of lean and fat red-eyed vireos. They found that lean vireos speed up their food finding rate, search in more habitats, and try different methods — flying after insects to catch them on the wing, for example. As the birds seek food, they must balance the need to gain weight quickly with keeping alert for predators and finding time to rest. One thing is clear: as they cope with meeting high energy demands, navigating in unfamiliar places, avoiding predators, and competing with other birds for food, stopping over during migration is a fast-paced and stressful business.

Migratory songbirds spend as little time as possible in the cheniers. Birds with enough fat to travel onward usually take off at the first nightfall when the weather permits; the others usually leave within a few days, as soon as they have built up enough fat reserves to continue. Every day that passes there will be more competition for nesting sites when they arrive on the breeding grounds and less time to build a nest and raise their young before they must depart for their long southward migration in fall.

On May 2, at the end of the day, the field crew closed the nets, weary but elated that they had recorded for history the measurements of so many songbirds on a day that surpassed anything they had ever seen. The next morning, with the winds dying back, it was still a busy day with 185 birds captured. Some already had shiny new bands—recaptured birds that had been banded in recent days. Although they were still trying to gain enough weight to travel again, many others had undoubtedly moved on. They were carrying the aluminum bands with numbers written down on the banders' data sheets, showing that they were here, in Johnsons Bayou, Louisiana, on May 2, 2004, as part of one of the biggest fallouts recorded in recent years. Some, such as the Tennessee warbler that hit Németh's window screen, may have survived only narrowly. Reenergized after their stop in the cheniers, they would now be flying north, part of an unstoppable mass migration, with hundreds of miles still ahead before their spring journey would be complete.

3
THE THRUSH CHASERS

Sweet bird! thy bower is ever green,
Thy sky is ever clear,
Thou hast no sorrow in thy song,
No Winter in thy year.

O could I fly, I'd fly with thee!
We'd make with joyful wing
Our annual visit o'er the globe,
Companions of the Spring!
— *Michael Bruce, from "Ode to the Cuckoo"*

IN SPRING, gray-cheeked thrushes travel some four thousand miles from the rainforests of South America to the tundra and taiga of Canada and Alaska. They migrate through the midwestern United States by the tens of thousands, flying above forests, lakes, farms, and cities — but unless you listen for their high-pitched calls, you would never know they were passing by in the dark. Even after daybreak, when gray-cheeked thrushes have sifted down from the sky and into the trees, most people never know they are there, unless they glimpse one of these shy birds hopping among fallen leaves before disappearing into the brush.

On May 24, 1965, Richard Graber was waiting for thrushes in an orchard in Urbana, Illinois. At seven that morning, he had strewn six mist nets among the shrubs. Already he had caught a Canada warbler, a red-eyed vireo, a Wilson's warbler, and an ovenbird, but they were too small for his purposes and he had let them all go. It wasn't until noon that he finally caught what he was waiting for — a gray-cheeked thrush, a bird heavy and strong enough to carry a miniature radio transmitter on its back while it flew. If his plan worked, he would ascend with the thrush by airplane that night and follow the bird wherever it migrated by listening for the pulse of the transmitter.

An ornithologist with the Illinois Natural History Survey, Graber had studied bird migration using radar images. He had audio recorded and quantified the birds' calls as they passed overhead. He had traveled to television towers to pick up hundreds of dead birds that had struck the towers at night and fallen to the ground — tangible evidence of mass migrations. But clues such as these didn't tell Graber what he really wanted to know. How did a bird decide when to take flight, where to go, how fast to fly? Migrating with the birds was the only way to find out.

Using radio transmitters, biologists had tracked large birds such as ducks and homing pigeons in flight, but the transmitters had been too heavy for songbirds. The lightest ones had been about half an ounce, more than the weight of an entire chickadee. To track a large songbird such as a thrush, someone would have to package the technology into a neat bundle weighing less than one tenth of an ounce, about as much as a dime. Fortunately, Graber thought he knew just the right person to do it — his friend Bill Cochran, an electrical engineer.

After a few weeks of work, Cochran presented Graber with the transmitter in January 1965. By late May, they had released eighteen thrushes with transmitters glued to their backs. A few of the transmitters had fallen off before the birds ever took off. But when the transmitters stayed on, Graber and Cochran were able to detect the thrushes from as far away as a few miles if the bird was on the ground, or 25 miles if the bird was in the air. They tracked one Swainson's thrush for more than 200 miles into Missouri, and another for more than 350 miles into Minnesota.

On May 24, as Graber gently untangled the gray-cheeked thrush from the net, he noticed faint spots on the grayish-brown wings, the marks of a young bird that had hatched the previous summer. In the first months of its life, the thrush had flown perhaps to Colombia or Venezuela for the winter; now it was on its way back north, having completed seven thousand miles of its eight-thousand-mile round-trip. Graber held the softly feathered thrush in one hand and used his other hand to dab a spot of glue on the button-sized transmitter with its foot-long antenna. He pressed the transmitter onto the bird's back, then released the thrush into the trees. He retreated to his truck parked in an adjacent field and tuned the receiver to the radio's frequency. "*Cheerp, cheerp, cheerp,*" the signal came through. All afternoon, the bird remained in the woodlot. When it moved to preen itself or hop about in search of food, the radio signal fluctuated as the thin wire antenna vibrated on its back.

Before dusk, Graber drove to the nearby Illini airport, where he waited nervously with pilot Jim Taylor, wondering whether the thrush would fly that night. At seven fifty-five, the thrush took off, its signal increasing in strength as it rose into the air. Minutes later, Graber and Taylor were airborne, listening to the thrush's signal through headphones. Below the airplane, the lights of small towns glittered in the dark and the stars shone down from above. Taylor kept the airplane within a mile of the flying thrush as it headed northeast at about fifty miles per hour. "The experience was unique," Graber later wrote. "A thrush was guiding us to some unknown destination."[1]

At eight forty-nine, the thrush was passing over Thawville, Illinois, still traveling at fifty miles per hour and heading straight for Chicago. Mile after mile, Graber and Taylor followed the bird's signal in the darkness. Then, as they approached the south edge of Chicago, they began having trouble hearing the signal because of engine noise from the airplane. At ten fifteen the noise drowned out the signal altogether.

Quickly, Graber marked a line on the air chart that showed which way the thrush was heading. He calculated that if the bird kept flying in the same direction at the same speed, it would cross north of Evanston, Illinois, at ten thirty-seven. Taylor headed the airplane toward Evanston, then circled just north of the town while they waited. At ten thirty-seven, they picked up the bird's signal, right on schedule. As they flew northward again in pursuit of the thrush, the scattered lights below suddenly dropped off into pitch blackness. The bird was leading them out over Lake Michigan.

"My hopes sank," Graber wrote. "Unless it changed its course, it would fly virtually the length of Lake Michigan, almost 250 miles over the open water—this in addition to the 140 miles it flew before reaching the lake. Nearly 400 miles and more than 8 hours of flying with no chance of rest."[2]

The airplane's fuel was low—too low to make the flight that the bird would have to complete on just a few grams of fat. Reluctantly, Graber and Taylor swung back toward the mainland. "Where that lone, delicate bird went, we could not risk going," Graber wrote. "In that moment, as the signal faded in the darkness, I felt overwhelming admiration for that bird . . . Thinking of the transmitter, I flinched. What right had I to add almost three grams of useless burden to the hardships this delicate creature faced? If only I could take it all back. Let the bird keep its secrets, only let it be safe. That was my fervent wish. Yet nothing could change what was already started."

Based on the bird's heading and speed, Graber calculated that the thrush should cross Sturgeon Bay, Wisconsin, around two forty A.M. They could try to intercept the bird there after refueling. Meanwhile, the thrush was flying directly into the path of a severe thunderstorm. Already the sky had clouded over and rain was pelting the airplane's windows. After Taylor landed the plane in Green Bay, Wisconsin, Graber stepped out of the airplane for a moment to look up at the sky. As cold raindrops fell on his face, he could hear the high-pitched calls of thrushes and warblers flying overhead. A few minutes later, he and Taylor were airborne once more, racing to find their gray-cheeked thrush among thousands of songbirds migrating northward.

When they reached Lake Michigan again, bolts of lightning were shooting down over the water. Graber thought about the solitary bird, warm and filled with life as it winged its way through the stormy sky. He imagined the rain-drenched thrush becoming too exhausted to fly and fluttering down into the dark water. It was such a painful scenario to contemplate that he tried to banish the thought.

Tense with anticipation, Graber and Taylor approached Sturgeon Bay around two thirty A.M., within five miles of where Graber calculated that the thrush would pass by. If their timing was right and the bird was within twenty-five miles, they should hear the signal. They circled the area, listening for the signal in between crashes of thunder. Ten minutes went by as they listened in vain. "There's no hope, Jim," Graber finally said. "Let's go back."[3] But Taylor kept circling. The minutes slowly ticked by until suddenly, at two forty-eight, they both heard the familiar beeping coming in over the headphones. Taylor circled again, then flew toward the signal until Graber was convinced—the bird was still flying on the same heading that it had ever since taking off from the orchard in Urbana.

"Neither of us actually spoke aloud, but there was a sigh of relief between us that said, 'Thank God! He made it,' " Graber wrote.[4] Since their last contact with the thrush, four hours and nine minutes earlier, the bird had traveled two hundred miles over the water and was still heading northward at about fifty miles per hour. If it kept going, it would reach Washington Island in forty more minutes, around daybreak. Yet Graber and Taylor never found out whether the thrush landed safely. The storm and the dense fog forced them to turn back as the thrush flew on.

To Graber, the thrush's flight had seemed something of a miracle, as staggering to his imagination as an astronaut's flight to the moon. He would never forget how the creature he had held in his hands at midday had flown more than eight hours that night, crossing four hundred miles. In cloudy skies without stars or the moon to help show the way, it had kept an unerringly straight path, never stopping to eat, drink, rest, or seek shelter from the storm. Before its long annual journey was ended, it would have to cross another one hundred miles over Lake Superior then onward into Canada. Many unanswered questions remained, but Graber hadn't liked interfering with the thrush's fate — not when it had already survived the odds of crossing so many perilous miles. "Let the bird keep its secrets, only let it be safe," he had written, and it seems that he meant it; he never tracked another thrush again.

IN THE SPRING OF 1966, Graber asked Cochran to continue the project for the Illinois Natural History Survey. Without the funds to track birds by airplane, Cochran decided to follow the thrushes by car and live out of his vehicle along the way to cut costs. He mounted an oversized antenna on top of his black Chevy 10 and cut a hole in the roof so that the antenna's pole could protrude into the passenger compartment. While the driver pursued the thrush, the backseat passenger could rotate the antenna, plot the bird's course on road maps, and help navigate the best routes. Fortunately, the roads in Illinois crisscrossed in an orderly grid, making it possible to pursue a thrush in a straight line if it traveled north, south, east, or west, or to zigzag diagonally if it headed off in directions in between.

It was a labor-intensive method, but tracking birds and devising new gadgets to study migration became Cochran's lifelong pursuit. In 1981, he designed a transmitter that pulsed with the movement of the birds' wings, allowing him to record how frequently the birds flapped their wings during takeoff, migration, and landing. In 1994, he attached a miniature microphone transmitter to the backs of birds so that he could study their vocalizations in flight. From 1965 to 2004, he tracked an assortment of birds across some 30,000 miles, requiring about 150,000 miles of actual ground travel to keep up and search for the signals after landing.

Cochran followed flocks of tundra swans, sandhill cranes, and Canada geese. He tracked two Cooper's hawks and five sharp-shinned hawks, one of which led him from Cedar Grove, Wisconsin, to Huntsville, Alabama in 1972. He followed two golden eagles, six merlins, a gyrfalcon, and twenty peregrine falcons. In 1975, he traveled with one peregrine from Little Suamico, Wisconsin, to Tampico, Mexico. In 1992, Cochran and his son tracked a nighthawk from Mahomet, Illinois, to a town near Charleston, South Carolina, where they heard the transmitter's last signals fading out over the Atlantic Ocean. But mostly Cochran tracked thrushes — about sixteen wood thrushes and eighty Swainson's thrushes, hermit thrushes, gray-cheeked thrushes, and veeries.

In the process, Cochran gathered reams of data about

the behavior of migrating thrushes. The radio signals confirmed others' observations that, by daylight, thrushes seldom moved more than a few hundred yards from where they had landed. They spent the day resting, preening, and hopping about the underbrush in search of food. Cochran's data also showed that whether or not a thrush took off at night depended on how well its energy reserves could sustain a long flight. Usually only thrushes weighing slightly more than an ounce would resume migration; those weighing less than an ounce stayed longer, until they had gained enough fat.

Thrushes heavy enough to resume their journey took off just after sunset, usually on days when the air temperature had risen to at least sixty-nine degrees in the shade and wind speeds at takeoff were six miles per hour or less. These rules of thumb were so reliable that Cochran once tracked a thrush from Urbana to Chicago, then decided to drive two hours back home when a cold front blew in. For the next ten days, he waited comfortably at home, confident that the thrush would not take off until the forecast predicted high temperatures near sixty-nine degrees.

As long as the nighttime temperatures remained stable, the thrushes were undeterred by storms, even with heavy rain and lightning. Cochran followed numerous thrushes into thunderstorms; some birds even inexplicably abandoned their course to head straight toward areas with lightning. In one harrowing trip, Cochran followed a thrush by airplane into a lightning storm. As sparks jumped off the airplane's antenna, Cochran radioed his will to a crew on the ground, but the thrush flew steadily onward through soaking rain, loud cracks of thunder, and flashes of lightning.

In contrast, a subtle change in temperature was often enough to cause the thrushes to land immediately. Although the birds flew on nights as cold as thirty-nine degrees, they stopped as soon as they encountered a cold front, even if the drop in temperature was as slight as three degrees. One night Cochran had tracked a Swainson's thrush for 168 miles from Iowa to Minnesota when, near midnight, his Chevy ran low on fuel. He sped ahead to buy gas in the nearest town, then waited for the bird to catch up, its signal passing overhead.

After waiting with no sign of the thrush, Cochran noticed that the night air felt colder and the winds had shifted, gusting at twenty miles per hour from the north. If the thrush had landed because it had encountered a cold front, he might be able to calculate where the bird would have gone down, rather than relying on pure luck while searching hundreds of square miles. At a nearby airport, Cochran found a meteorologist on late-night duty. Using information from the meteorologist about when the wind changed and the speed of the cold front, along with the thrush's last known location, heading, and speed, Cochran calculated that the bird would have landed near Simpson, Minnesota. After driving there, he was astounded to find that the thrush had landed within half a mile of where he had drawn the spot on the map.

Unless they encountered a cold front, the thrushes usually continued flying all night, or until their body weight dropped to the level of the previous morning. Even on dark, cloudy nights, the thrushes usually managed to land in wooded areas. Sometimes when the weather forced them down, they ended up in the middle of cornfields. A few landed in the midst of Chicago or other major cities and spent the day in the trees of urban neighborhoods.

Cochran also documented how individual thrushes traveled along different headings, but each bird appeared to maintain its own constant heading throughout its journey. The birds could be pushed off course by strong winds — some of them were blown due east or west — but usually the deviations were slight and averaged out over many flights. How did they stay on course? By studying captive birds, ornithologists had discovered that songbirds could use several kinds of clues for orientation — landmarks, stars, the position of the sun, patterns of polarized light, and even the earth's magnetic field. Yet little was known about how the birds used these cues in free flight. Cochran reasoned that the thrushes must use a magnetic compass, since they could stay on course even on cloudy nights. But they must also use some other cue; otherwise, they could become confused when crossing the magnetic equator or flying through anomalies in the earth's magnetic field.

To test this idea, Cochran conducted experiments by exposing thrushes to an altered magnetic field before takeoff. In 1978, 1979, and 1984, he captured eight thrushes, put them in cages during sunset, and shifted the magnetic field around them to the east. He managed to follow two of the birds for two flights in a row. He discovered that on the first night, the birds flew westward rather than flying north. On subsequent nights they migrated northward again. This meant that the birds were indeed using a magnetic compass — the altered field had caused them to orient in the wrong direction. Cochran concluded that they corrected their path on the next flight by using the position of the setting sun or the pattern of polarized light near sunset to calibrate the compass. In 1986, Cochran submitted the results to the journal *Science*. The editors wrote back with a rejection letter explaining that the week's issue was already full; there was no need to bother submitting the paper for reconsideration.

It wasn't the first cool response that Cochran had received from the scientific community. In 1973, he tracked a Swainson's thrush over the course of seven nights and 940 miles—an unprecedented distance that is still unrivaled more than thirty years later by any other attempt to track songbirds. The journal *Animal Behaviour* published the study as a minor scientific note and refused to include a map that showed the bird's dramatic progress from Illinois to Manitoba, Canada.

Perhaps Cochran's technique of chasing thrushes at night was too unconventional. To some, the data may have seemed unsatisfyingly anecdotal—after all, his journeys revealed stories about individual thrushes, not to be equated with a general truth about songbird migration. To others, the birds' behavior may have been interesting but tainted. Were the birds really migrating freely if they had been burdened by an unnatural load that was glued and harnessed to their backs? It didn't help that Cochran was an outsider, an engineer who was quick to admit he still didn't know much about birds—even though he probably knew more about the migratory behavior of North American thrushes than anybody else.

For decades, the results of most of Cochran's work remained unknown to the scientific community. "People knew that there was this weird ingenious person out there, sort of a gypsy driving behind a thrush," says Martin Wikelski, a biologist at Princeton University. "But they didn't really understand how far reaching his ideas and techniques were."[5] Virtually everyone who studied long-distance songbird migration used other approaches. Some captured and banded thousands of birds, hoping to recover a few of them on their wintering grounds or along their migratory routes. Others used radar to study mass migrations or tested caged birds under various conditions to find out how they know which way to fly. In later years, new molecular techniques made it possible for researchers to narrow down which region of the continent the migratory birds had inhabited, by analyzing blood and feather samples for specific elemental isotopes characteristic of those regions. With less success, they also attempted to use DNA to determine whether birds breeding in one location could be linked to those wintering at another location.

IT WAS ONLY IN 1999, after Cochran had been tracking thrushes for thirty-four years, that someone else adopted his techniques. Wikelski, a young biologist then at the University of Illinois, had become excited about the possibility of measuring how much energy birds expended in migratory flight. Before releasing a thrush with a transmitter, he planned to inject it with doubly labeled water, a form of water with distinctive hydrogen and oxygen isotopes that is used to measure energy expenditure in humans and other animals. After following the thrushes by car, he would recapture them and take blood samples to compare with the samples taken before takeoff.

Wikelski and his collaborators succeeded in recapturing

twelve thrushes and measuring their energy expenditure. The results were counterintuitive: on a journey north in spring, thrushes expended less energy on actual flight than they did while resting and refueling during stopover periods. Feeding a large brood of young required more energy per day than the migratory flights that many had previously assumed were so costly. A thrush could complete a three-thousand-mile trip from Panama to Canada in eighteen flights over the course of forty-two days, averaging about 158 miles, or 4.6 hours per flight, and expending about 0.3 calories per mile, including flights and stopovers. Wikelski and his collaborators published their findings in the prestigious journal *Nature,* and stories of their nighttime pursuits and run-ins with small-town police made international news. The study brought new attention to radio tracking just as Cochran was getting ready to retire his tracking vehicle and do something else for a while.

Wikelski, energetic and enthusiastic, managed to persuade Cochran not to give up on the field work just yet. Together, with colleague Henrik Mouritsen, they revived the magnetic orientation experiments, increasing the sample size to make the study more robust. This time, *Science* made room for the results. As Cochran had found earlier, the birds exposed to the magnetic field initially flew in the wrong direction but corrected their path the next night. "In the morning, shortly before they land, they see the sun and realize they have made a mistake," Wikelski told reporters. "You can see them turn around 90 degrees." The birds were relying on the location of the sunset to determine which way to fly and maintained the heading in darkness using a magnetic compass at night. "It is the simplest and most foolproof orientation mechanism we can imagine," Wikelski said.[6]

Cochran and Wikelski also began using radio transmitters that Cochran had designed to detect birds' heartbeats and wingbeats as they flew. For the first time they recorded how quickly a songbird's heart pumped as it flew through the sky — about 840 beats per minute. They simultaneously recorded the birds' wingbeats, finding that the thrushes beat their wings 600 to 780 times per minute — about 3.2 million wingbeats for its entire continental flight.

In 2005, Cochran and Wikelski published for the first time a summary of major findings from the past three decades of tracking thrushes, giving scientists a rich new source of information about songbird migration. The data from the flights of even a single thrush are still unmatched by any other method of study. "The seven nights that Bill followed the thrush to Canada is still the best information we have on migrating songbirds," Wikelski said. "There's nothing comparable out there. The problem is that to get this kind of data, you have to work like crazy."

When Cochran and his assistant, Chip Welling, followed the Swainson's thrush for seven nights in 1973, they nearly lost the bird three times. In Iowa, the signal vanished when the thrush took a rare daytime excursion while Cochran and Welling were eating breakfast. In Minnesota, they lost the bird when they ran low on fuel and had to look for a gas station. Farther north in Minnesota, although they were driving below the speed limit, a police officer stopped their suspicious-looking vehicle, cited them for speeding, and ultimately caused them to lose the bird again. In all cases, they relocated the thrush with a combination of luck, wits, and dogged determination.

They followed the thrush all the way to Canada, but a closed border station — and a broken radiator — prevented them from continuing the chase. Near four in the morning on the last leg of their chase, Cochran and Welling had to sit in their Chevy in North Dakota, their path blocked by Styrofoam barriers at a border station, as the thrush flew onward in the dark. It was excruciating, sitting at the empty station, knowing they could push the barriers out of the way and drive on. More than thirty years later, Cochran said he wishes that they had risked it, even though they would have lost the thrush if they had been arrested.

As it was, they crossed into Canada when the station opened later that morning, but within hours they were forced to stop on a rural road when the Chevy's radiator boiled over and ran dry. "We found a ditch that had water in it and we had some paper cups," Cochran said. "It took one heck of a long time to fill that radiator with trips down to the ditch." They never found the thrush again. "It was sad to leave," Cochran says. "For several nights thereafter, before falling asleep, I would hear the beeping sound. I'd been listening to it off and on for so many days that as I went to sleep I'd hear it even though it was no longer there."[7]

IN THE FUTURE, a new generation of radio transmitters or satellite tracking from space could make challenges of the road obsolete. Scientists at the Cornell Laboratory of Ornithology are developing radio transmitters that use a clever combination of tricks to track birds automatically over greater distances. Conventional radio transmitters weigh about ten times less than those of forty years ago, but one of their biggest drawbacks is their short battery life. They function for only seven to ten days before the batteries wear out, making it difficult to follow a migratory songbird very far. Now engineers at the Lab of Ornithology are designing transmitters that can be programmed to turn off when the bird is inactive, rather than beeping incessantly for twenty-four hours a day. Researchers will also be able to specify when the transmitter should produce a signal to provide an update on the bird's location. These energy-saving measures will extend the precious battery life to years.

Instead of following birds closely from vehicles on the ground, researchers could implement technology used in cellular telephone networks to track birds automatically. A network of receiving stations would detect signals from the transmitters and identify the bird's position. If the bird flies out of range of receiving stations, the transmitter could collect and store information about the bird's location and relay the data when the bird returns within range of a station.

The transmitters will record information about where the birds are by using light sensors that detect the onset of dawn and dusk. Comparing those times against a standard clock, such as Greenwich Mean Time, could reveal where the bird is as it travels over the surface of the earth. Heart rate, temperature, or humidity detectors could also be included onboard a transmitter. Researchers could put transmitters on migratory birds before they leave in summer and collect information about their migratory route, environment, and physiology when the birds return from the tropics in spring.

Unfortunately, equipping songbirds with transmitters is time-consuming, costly, and often inefficient since many songbirds have mortality rates of about 50 percent annually. To increase the chances of gathering data even from birds that never make it back, researchers from the Lab of Ornithology and the Cornell College of Engineering are designing transmitters that can exchange information with one another. When two migrants approach within a certain range, their transmitters could relay data to one another. If one bird perishes, researchers could still gather data about both birds from the surviving bird's transmitter. The transmitters could even be used to exchange information from different kinds of animals. Transmitters carried by whales in the ocean could exchange data with seabirds migrating overhead. When the seabirds return to their breeding colony, researchers could pick up information about where the birds and the whales have been.

Martin Wikelski, meanwhile, is working with an international team of scientists to track songbirds from space. The birds would carry transmitters that emit signals picked up by space satellites and report to a computer back on Earth. Finding out where the bird traveled overnight would be as simple as waking up in the morning and taking a look at the computer screen.

Although satellite transmitters are still too heavy to use on songbirds, scientists have used the technique to track larger birds. They have documented a peregrine falcon migrating from Alberta, Canada, to Mazatlán, Mexico. They have tracked a spectacled eider from coastal Beringia to previously unknown wintering quarters on crevices in pack ice in the Bering Sea. A Swainson's hawk with a satellite transmitter traveled from California to the pampas of Argentina, revealing a new wintering location where thousands of Swainson's hawks were gathered. Among them were hawks that had been banded in California, Colorado, and Saskatchewan. Satellite transmitters have finally made it possible to study the entire migratory routes of individual birds.

The lightest transmitters able to reach satellites weigh eighteen grams — about ten times too heavy for a thrush to carry. However, the minute signals from radio transmitters could be detected using radio telescopes that have been trained on the distant universe to pick up faint signals from outer space. If a radio telescope were sent to space in order to listen to sounds from the earth, it could detect weak signals from a songbird carrying a transmitter on its back. "For radio astronomers, it's bread and butter technology," Wikelski said.

The biggest obstacle is not technological, Wikelski argues. Rather, it's in convincing NASA to divert some of its attention from projects on Mars and other places in the universe to focus on the planet right beneath our feet. "We need about $30 million — a ridiculously cheap sum for a project of that scale. I would bet more than that is spent annually on isotopic analyses, DNA analyses, banding of birds, and all that, trying to figure out the connectivity of birds. We would gain a tremendous amount of information if, in addition to using these traditional methods, we could detect the movements of birds directly from space."

Understanding the connectivity of birds — the connections between all the places they inhabit throughout the year — has tremendous implications for the conservation of birds. First, there is the obvious importance of simply knowing where the birds go. When Swainson's hawks were tracked by satellite to Argentina, biologists found that thousands of the hawks were dying on the pampas from an agricultural pesticide intended to kill grasshoppers. During the winter of 1996 alone, scientists estimated that as many as twenty thousand Swainson's hawks were killed in a portion of La Pampa, a staggering loss that, if continued, would endanger the entire species in just a few years. The findings mobilized an international campaign to discourage the use of the pesticide and raise awareness about its harmful effects. In addition to learning more about where birds winter, tracking them for their entire migratory route would help identify which areas need to be protected to ensure they can reach their destinations.

New methods of tracking would open the way to exploring the movements of birds around the globe. "I would love to know how bar-tailed godwits go from Alaska to New Zealand," Wikelski said. "What about the warblers that supposedly fly out over Boston and come down in the Caribbean — is that what they really do? Then there's migration within the tropics — we have absolutely no clue what's happening there. There are two months of the year when millions of birds from Europe disappear in Africa. Nobody knows where they go. All of those questions are out there to solve."

Perhaps Wikelski and others will be able to track birds someday without worrying about lightning storms, gas stations, police run-ins, political borders, and broken radiators. They could track birds across intercontinental distances by going to sleep at night and letting the technology do the

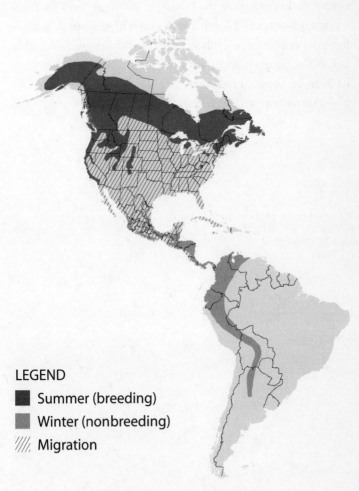

Summer (breeding)
Winter (nonbreeding)
Migration

Swainson's thrush migration map.

work of tracking the birds. Yet Wikelski suspects there will always be an element of road travel because it isn't enough simply to know where the birds go. A thrush may fly from a certain forest in Colombia to a patch of taiga in Alaska — but how does it decide where to stop at either end, and what are the consequences of its choices? Someone will always have to be there, traveling where the birds travel, to really understand how and why they do it.

4
WITNESSING THE SPRING SPECTACLE

Why read a book when there are birds
Printing clear and breezy words
Upon the cloud's white pages? When
A busy robin and a wren
Are syllables of ecstasy!
A line of swallows on a tree,
Or wire, is a sentence, long
And sweeping. A flying flock's a strong
Paragraph, while in the air
Is quilled elaborately and rare
Illumined manuscript in gold
And green. And say, what book can hold
More fascination and delight
Than birds in migratory flight?
—*Collette M. Burns, "Why Read a Book?"*

Spring is a spectacular time to watch migratory songbirds as they pass through North America in dazzling color, abundance, and diversity. On any given day, however, what birders actually see may be disappointing or amazing, depending on winds, weather, geography, and the calendar date. Though there is always an element of the unpredictable, a few simple rules of thumb can help explain why some areas are hot spots and can help forecast whether many birds or few will show up on a particular day.

Songbirds migrate across a broad front, so hot spots tend to have geographical features that concentrate migrants in those areas. Birds sometimes land in legendary numbers near geographical barriers to migration, such as the Gulf of Mexico and the Great Lakes. Rain, headwinds, or fog can precipitate fallouts of birds reluctant to cross water or that are too exhausted to fly any farther after reaching the other side. Peninsulas are especially good places to look for migrants. Depending on where they are located, they may funnel birds crossing over land toward water or beckon to migrants as they reach the other side. Astounding numbers of songbirds can also become concentrated on islands, whether masses of land surrounded by water or welcoming patches of greenery surrounded by open deserts, agricultural fields, or urban sprawl.

Extensive forests always attract migrants and are essential for providing good stopover habitats, but some of the most exciting bird watching may be in small, isolated patches that force the birds into a restricted area. In the Cayuga Basin near Ithaca, New York, for example, birders have their pick of forests and lake shores for watching migrants, but a favorite spot is in the Hawthorn Orchard, a thirteen-acre area surrounded by homes, tennis courts, an equestrian center, and a shopping plaza. When the hawthorn buds open in May, as many as forty-three species of migratory songbirds crowd the trees to feed on insects.

Timing also has a key influence on the kinds and numbers of songbirds passing through. The first waves of birds in March usually include short-distance migrants coming from the southern United States and Mexico, such as red-winged blackbirds and eastern bluebirds. April usually brings the first warbler species. The males often arrive first, risking early spring weather for the chance to claim the best territories. Spring migration peaks during roughly the same few weeks each year—for example, the last two weeks of April along the Gulf Coast and early to mid-May in the Northeast and Midwest. During peak periods, birders have the best chance of seeing the greatest number of birds and species, but migration continues into late May, bringing greater numbers of latecomers such as blackpoll warblers and mourning warblers than in the earlier weeks.

Weather is the wild card, determining when and from where millions of migrants may take off, land, and remain grounded. In extended periods of favorable weather—with tailwinds, warm temperatures, and lack of rain—most birds take off each night and migrate northward steadily. By morning they have diffused across the landscape. Because they are encountered in lower density, it may seem as if migration is "slow," when in fact conditions were optimal for large numbers of migrants. On those days, however, bird watching is good in wooded areas surrounded by inhospitable areas such as cities, marshlands, and agricultural fields. As birds arrive on south winds and find few good choices about where to land, many of them will drop down into woodlots, forested cemeteries, and urban parks.

In areas where birds have better choices about where to land, such as in large tracts of forest, the best days for bird watching occur when weather stops birds from migrating. Cold fronts bring northerly winds and rain, causing birds to land and remain grounded until the weather improves. When cold fronts to the south linger for several days, birds may pile in, then take off only when the temperatures warm and they can take advantage of southerly tailwinds again.

When optimal geography, timing, and certain weather conditions converge, they create bird-watching events of a lifetime. One such fallout occurred on May 4, 2002, during peak migration in the Northeast. The locale was the Forsythe National Wildlife Refuge, just north of Cape May, New Jersey, a peninsula on a major migratory flyway along the Atlantic Coast and a world-renowned hot spot for migratory birds. For two straight nights, as birds traveled up the coast, powerful southeasterly winds had pushed them out over the ocean. Early in the morning on May 4, the winds shifted and the birds headed for land. Birders standing on a dike in the wildlife refuge were amazed to see songbirds pouring in from over the Atlantic at eye level or below. The birds passed so closely that they could have been caught with baseball gloves, according to Cornell Laboratory of Ornithology director John Fitzpatrick, who witnessed the fallout with ornithologist Kevin McGowan. "Stepping out of the car, Kevin and I began pointing and exclaiming with child-like delight as dozens of scarlet tanagers, Baltimore and orchard orioles, magnolia and black-and-white warblers

zipped by," Fitzpatrick said. "With each approaching pha-
lanx, though, we grew more silent. This was a spectacle neither of us had ever witnessed before. Giddy wonderment
was giving way to awe."[1] It was a life-changing glimpse,
Fitzpatrick said, of fantastic journeys that normally produce smaller spectacles over our heads, at night, or out at
sea.

Although fallouts are not always predictable, some birders scrutinize weather and radar images to predict the magnitude of migration. During 2000–2001, a project called
BirdCast posted spring migration forecasts through Bird-
Source, an interactive Web site that allowed birders and scientists to work together in collecting data about spring
migration. The bird forecasts were based on a model developed by Sidney Gauthreaux and collaborators at the Clemson University Radar Ornithology Laboratory. Researchers
analyzed how well their forecasts held up by examining the
radar images corresponding to the forecast periods and by
asking birders to report what they saw in the field. BirdCast
no longer operates, but you can formulate your own predictions using a few hints, explained by Andrew Farnsworth,
the "BirdCaster" who formulated each day's predictions
and analyzed their success. Note that these predictors work
better in the East than they do in the West, where songbird
migration is more protracted and diffuse.

1. In the evening, around seven or eight P.M., look at
   continental weather maps such as those shown on
   the Weather Channel or on the Weather Channel
   Web site, http://www.weather.com. You can also
   find good weather maps on the Weather Analysis

Page from the College of DuPage, http://weather
.cod.edu/analysis/analysis.sfccon.html, or Real-
Time Weather Data from the National Center for
Atmospheric Research, http://www.rap.ucar
.edu/weather/model. These maps show
temperatures, high and low pressure systems,
and precipitation.

2. Identify the high pressure and low pressure weather
systems in your area to determine which way the
winds are flowing. Air circulates in a clockwise
direction around high pressure systems and
counterclockwise around low pressure systems.
This could tell you, for example, that a high
pressure system to the east will bring southerly
winds, probably bringing warm and moist air
favorable for migration in spring. A low pressure
system to the east will bring northerly winds with
cool and dry air that could slow or stop migrants.

3. Look for areas with precipitation or low cloud
ceilings. Heavy rain, fog, and poor visibility
conditions will often stop migrants, particularly if
the conditions are so widespread that birds have
difficulty flying around those areas. If your location
is near the southern edge of a rain system, you may
see many migrants that have landed by morning.

4. To get a sense of the magnitude of the migration,
look at radar maps such as those at the College of
DuPage Web site, http://weather.cod.edu/analysis/
analysis.radar.html. A good time to look at radar
images is from two to four hours after sunset,
when the highest densities of birds are typically

aloft. Birds detected on radar tend to appear as colored pixels with a stippled look in radar base reflectivity images, rather than the blocky chunks of pixels associated with precipitation. Radial velocity images can also aid in identifying whether the radar detections are birds. These images are color coded to show the speed and direction in which the birds are traveling. You can recognize birds by looking for targets that, according to the color-coded images, are traveling about ten to fifteen knots above the wind speed or in a direction that differs by more than thirty degrees or so from the wind direction. You can find local wind direction and speed data from your local airport, weather station, or news bureau.

The base reflectivity image on page 73 is a black-and-white version of a color-coded radar image from Grand Rapids, Michigan, on the evening of May 20, 2001. The large circle in the center of the image is a mass of migrating birds. In the lower left corner, a small elongated block indicates rain. In the color image, different hues correspond to the density of objects that the radar encounters. In this case, the highest concentrations of birds are in the small circular areas within the large circle; the lighter areas around the edge of the circle indicate lower densities. The radial velocity image (not shown here) indicated that the birds were traveling to the north-northwest at about twenty knots. The outline of Lake Michigan just to the left of the circle shows that birds were crossing the lake shortly after they took off at nightfall. A radar image from this same location later on in the night would show more birds migrating over the lake.

With practice, you can use radar images to guess whether few or many migrants are aloft and headed your way. For more detailed instructions on how to interpret radar images, visit the Clemson University Radar Ornithology Laboratory at http://virtual.clemson.edu/groups/birdrad.

5. Make a forecast. Are there many birds aloft? If so, southerly winds and ideal flying conditions will probably keep birds moving throughout the night. Along the Gulf Coast, birds will continue flying inland in favorable conditions with warm, moist, southerly winds. At inland locations north of the coast, birds will probably be relatively dispersed when they land in the middle of the day. However, in areas where bird have few choices about where to land, densities could be high in small wooded patches, such as in urban parks. If many birds are aloft, you could get a fallout of birds the next day if the southern edge of a cold front is positioned over your area with rain or poor visibility. What if there are few birds downwind, or if a cold front has just passed through and the winds are unfavorable for migration? You might choose to sleep in the next morning, but if a cold front to the south is holding migrants back, be ready to catch a big arrival as soon as the next warm front comes in.

6. Jotting down your predictions and seeing how well

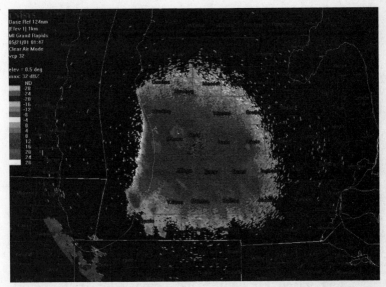

This radar map of Grand Rapids, Michigan, shows birds (large circular area) approaching and crossing Lake Michigan (oblong area to the left of most of the birds) shortly after they took off at nightfall. A radar image from the same location later that night showed more birds migrating over the lake.

they hold up the next day in different locations can be a fun way to learn how geography, local and regional weather conditions, timing, and sheer luck influence where and when we see songbirds in migration.

Hot Spots for Spring Songbird Migration

GULF COAST

The Gulf Coast is one of the most exciting regions to watch migrant songbirds in spring. Hot spots include High Island and South Padre Island in Texas, Grand Isle in Louisiana, Fort Pickens in Florida, and Dauphin Island and Fort Morgan in Alabama.

Some of the continent's most dramatic fallouts occur along the Gulf Coast during stormy weather, especially in heavy rains and north winds. Migrants that have battled headwinds during their six-hundred-mile journey across the Gulf may be so exhausted that they put down at the first sight of land. Fair weather and south winds make for notoriously slow days for spring bird watching on the Gulf Coast, however, since migrants take advantage of favorable winds to continue flying farther inland. Bird watchers congregate at Gulf Coast hot spots hoping to see a fallout, but feeling guilty, too, that they've come to marvel at events created by adverse weather conditions dangerous for migrant birds.

Fort Morgan, Alabama

For a special look at migratory songbirds as they arrive from across the Gulf, visit the banding station run by the

Hummer/Bird Study Group at Fort Morgan, Alabama. Led by Bob and Martha Sargent, the station is uniquely geared toward education and outreach. About three thousand visitors come during a two-week period each spring to learn about migration and to watch as volunteers capture, band, measure, and release hundreds of colorful songbirds.

As many as ninety species may be captured during the two-week period, including rarities such as black-whiskered vireos, a colorful array of warblers, and crowd-pleasing painted buntings in bright blue, red, green, and black. Beautiful orchard orioles, chestnut and black, pass through in numbers, and indigo buntings sometimes arrive in flocks, scattered on the ground in brilliant blues.

The number of birds captured depends on the weather, ranging from as few as a dozen birds per day to more than one thousand. Whatever the conditions, a visit is worthwhile. On busy days, extra volunteers are on hand to answer questions as the banders work without rest, handling a different bird every thirty seconds. On slower days, banders may give informal lectures about bird migration, behavior, identification, and habitat needs. Visitors have the opportunity to hold and release healthy birds after they have been banded.

WHEN TO GO: The banding station is typically open from dawn to midafternoon during the first and second weeks of April and second and third weeks of October. Admission is charged at Fort Morgan, but visits to the banding site are free.

NEAREST BIG CITY: Fort Morgan is located about two hours from Mobile.

SPECIAL ACTIVITIES: During the hour before dawn when the nets are open, banders sometimes catch nocturnal birds such as whip-poor-wills, chuck-will's-widows, and common nighthawks. Visitors are welcome to bring a flashlight and wait with the banding crew before dawn to see what turns up.

NEARBY SONGBIRD HOT SPOTS: At the banding station, pick up a free booklet describing birding hot spots on the Alabama Coastal Birding Trail. Be sure to visit Dauphin Island, just a forty-five-minute ferry ride away from Fort Morgan. Like Fort Morgan, Dauphin Island is known for spectacular fallouts in storms with north winds or rain.

FOR MORE INFORMATION: Bob Sargent: (205) 681-2888; http://www.hummingbirdsplus.org.

RECOMMENDED READING: "The Alabama Coastal Birding Trail." Pick up a free copy of this booklet at the Fort Morgan banding station or call the Alabama Gulf Coast Convention and Visitors Bureau: (800) 745-SAND; http://www.alabamacoastalbirdingtrail.com.

DESERT OASIS

Surrounded by immense stretches of desert, oases and riparian areas attract numerous songbirds in spring. Although the concentration of birds doesn't approach that of Gulf Coast and Great Lakes fallouts, birding in the desert has its own rewards. Near the shade and quiet of an oasis, you can see a variety of colorful birds as they linger in starkly beautiful surroundings. Desert locales known for spring migration include Malheur National Wildlife Refuge in Oregon, the San Pedro Riparian Area in Arizona, and Joshua Tree National Park and Anza-Borrego Desert State Park in California.

Yaqui Well, Anza-Borrego Desert State Park, California

A desert oasis renowned for spring bird watching is Yaqui Well in Anza-Borrego Desert State Park. For easy access, you can drive close to the well from a dirt road or hike a more scenic route from the Tamarisk Grove campground. On the 0.7-mile trail, you might hear the varied, musical notes of a California thrasher, encounter a clucking covey of California quail, or watch as a western tanager flies above the scrub, a fireball of yellow, black, and red. Yaqui Well itself is a small seep overgrown with cattails and surrounded by mesquite, described by wildlife biologist Paul Jorgensen as "a postage stamp riparian area."

The best way to watch for migrants is to sit for a morning and witness the surprising diversity of birds coming in to this patch of greenery. Nesting birds include Bell's vireo, verdin, cactus wren, phainopepla, black-throated sparrow, black-headed grosbeak, and Scott's oriole. Migrants such as lazuli bunting, Lawrence's goldfinch, and MacGillivray's warbler also stop through. At least fifteen species of warblers have been recorded in the park, including Lucy's, black-throated gray, and Townsend's.

Though Yaqui Well offers some of the best spring birding in the park, there are more than 600,000 acres and 100 miles of hiking and riding trails to explore. The 1.5-mile trail at Borrego Palm Canyon leads past an alluvial fan and along an intermittent stream to a pool and palm oasis good for migrants. Another popular bird watching site is in Agua Caliente County Park, where a trail leads to Squaw Pond, a small spring-fed oasis overgrown with reeds and shaded by willows and a fan palm tree.

WHEN TO GO: The last two weeks of April and first week of May are the best times to see migrant songbirds. Come prepared with water, hat, sunscreen, and a portable chair if desired.

NEAREST BIG CITIES: Anza-Borrego is about two hours from San Diego, Riverside, and Palm Springs.

SPECIAL ACTIVITIES: The visitor center offers free presentations about birds and other wildlife, November through May. Park naturalists lead occasional bird walks. In March and April, hawk enthusiasts should stop at the intersection of Henderson Canyon Road and DiGiorgio Road. This location was recently discovered as the best area in California to see migrating Swainson's hawks, with as many as 5,200 recorded in spring.

NEARBY SONGBIRD HOT SPOT: Joshua Tree National Park, two and a half to three hours away, spans a beautiful transition zone between the Mojave and Colorado deserts. Check for migrants in the oases and at Barker Dam.

FOR MORE INFORMATION: Anza-Borrego State Park: (760) 767-5311; http://www.anzaborrego.statepark.org/.

RECOMMENDED READING: *Guide to Birds of the Anza-Borrego Desert,* by Barbara W. Massey. Borrego Spring, Calif: Anza-Borrego Natural History Association, 1998.

GREAT LAKES

Numerous places around the Great Lakes offer opportunities to see migrant songbirds where they concentrate before or after crossing water. Well-known locales include Long Point Bird Observatory on Lake Erie near Port Rowan, Ontario; Whitefish Point Bird Observatory in Paradise, Michigan, at the southern end of Lake Superior; and Braddock Bay Bird Observatory along Lake Ontario near Rochester, New York. Lake Erie has several birding hot spots, including Point Pelee National Park in Ontario and the Crane Creek State Park and Magee Marsh Wildlife Area, in Oak Harbor, Ohio.

Point Pelee National Park of Canada, Leamington, Ontario

As migratory songbirds cross the Great Lakes into Canada, thousands funnel into Point Pelee National Park, a legendary place for spring bird watching. During certain weather conditions—such as a night of south winds combined with thunderstorms or early morning fog—the park can overflow with warblers, vireos, tanagers, orioles, and other spring migrants seemingly "dripping from the trees." Depending on the timing when the weather hits, large numbers of certain species may appear, such as in 1979, when the trees filled with so many scarlet tanagers that they couldn't be counted. Forty-two warbler species have been recorded in the park. Even on days unremarkable by Point

Pelee's standards, visitors may see twenty different kinds of warblers in a day.

Point Pelee is a fun locale to explore, with numerous trails and exciting bird sightings at every turn. Within eight square miles it has dry forest, red cedar savanna, swamp forest, freshwater marsh, and sandy beaches—a variety of habitats that appeal to diverse bird species. The park is situated where the Atlantic and Mississippi flyways cross, bringing in many different species and allowing rarities from both the east and the west to occur at this centrally located site. The birds concentrate here because, after a night of migration over water, Point Pelee is the first land they see, a six-mile sandspit jutting into Lake Erie. Futhermore, the park is the largest remaining natural refuge for migratory birds, surrounded by a county dominated by agriculture and human development. And because three sides of the peninsula are surrounded by water and its cooling effects, trees leaf out four to five days later than nearby areas, giving bird watchers a longer window of time in which to see these concentrations of migrants.

In spring, the roads and trails are lined with clusters of bird watchers, heavily laden with spotting scopes, binoculars, and cameras on tripods. To see the day's new arrivals, visitors usually walk or take a shuttle from the interpretive center to the tip. Breeding warblers sometimes nest at knee level by the boardwalk, while others flit through the shrubs and trees, among them Blackburnian, Cape May, Canada, blue-winged, golden-winged, bay-breasted, and mourning. The Delaurier Trail is another prime location, where yellow-breasted chats breed and nesting yellow-billed cuckoos can be found among the willows. In Tilden's Woods, to the northeast of the interpretive center, benches on a boardwalk

invite visitors to wait and see what comes to them —
perhaps a close-up view of a prothonotary or a hooded
warbler, a long look at a scarlet tanager, or a surprise appearance by a varied thrush or Kentucky warbler.

WHEN TO GO: Peak songbird migration typically occurs
around the second weekend in May.

NEAREST BIG CITIES: Point Pelee is about 180 miles southwest of Toronto, and about 35 miles southeast of Detroit,
Michigan.

SPECIAL ACTIVITIES: The interpretive center offers daily
theater presentations about the park and research projects.
Special programs take place during the Festival of Birds in
May, including, films, talks by visiting lecturers, and hikes
led by expert birders.

NEARBY SONGBIRD HOT SPOTS: Rondeau Provincial Park is
about a one-hour drive away, near Blenheim, Ontario. Long
Point Bird Observatory near Port Rowan, Ontario (about
2.5 hours away), conducts mist-netting and banding sessions open to the public.

FOR MORE INFORMATION: Point Pelee National Park of
Canada: (519) 322-2365; http://www.pc.gc.ca/pn-np/on/
pelee.

RECOMMENDED READING: *A Birder's Guide to Point Pelee,*
by Tom Hince. Wheatley, Canada: Wild Rose Guest House,
1999.

METROPOLITAN OASIS

During migration, surprising numbers of songbirds stop over in large cities where cemeteries and parks offer swaths of greenery. Some of the most famous urban hot spots for spring and fall migration are the Lakefront Park and Sanctuary (also known as "The Migrant Trap") in Hammond, Indiana; Green Lawn Cemetery and Arboretum in Columbus, Ohio; Mount Auburn Cemetery in Boston, Massachusetts; and Central Park in New York City.

Central Park, New York City

Amid one of the largest metropolises in the world, Central Park includes 843 acres of meadows, woodlands, lakes, and other welcoming habitats for birds. Designated an Important Bird Area in 1998, Central Park is renowned for spring migration, when as many as thirty different warbler species have been observed in a single day. Year-round, 275 bird species have been recorded.

If you're birding in Central Park for the first time, consider joining one of the regularly scheduled bird walks offered by local groups (see "Special activities" below). Field ornithologist Starr Saphir leads bird walks to the Ramble and the North Woods, areas that attract spring migrants such as chestnut-sided, magnolia, black-throated blue, and Canada warblers; scarlet tanagers; Baltimore orioles; and red-eyed vireos. The edge habitats along Rowboat Lake are especially good places to check for migrants, Saphir says.

If you go bird watching on your own, arrange to go with friends or other birders, and be vigilant in secluded areas. Pick up a park map in the Dairy (midpark at Sixty-fifth Street) or at Belvedere Castle (midpark at Eighty-second

Street). A good place to start is at the Loeb Boathouse on the east side between Seventy-fourth and Seventy-fifth streets. There, you can check the "Bird Register" to find out where others have seen birds of note recently. To the west, you can enter the Ramble. The North Woods is a more secluded area with more mature forest and a stream.

WHEN TO GO: Spring migration peaks in May, but interesting migrants may be found from late February through the first week in June. Bird watching in fall is also rewarding.

SPECIAL ACTIVITIES: The Charles A. Dana Discovery Center on Harlem Meer has a visitor center and free community programs offered by the Central Park Conservancy. The following groups lead bird walks. Call or check their Web sites for information about schedules, fees, and locations.

Central Park Conservancy: (212) 310-6600; http://www.centralparknyc.org.

New York City Audubon Society: (212) 691-7483; http://www.nycaudubon.org

Urban Park Rangers: (866) 692-4295; http://www.nycgovparks.org/sub_about/parks_divisions/ urban_park_rangers/pd_ur.html

Linnaean Society of New York: http://www .linnaeannewyork.org

The Nature Conservancy: (212) 997-1880

NEARBY SONGBIRD HOT SPOTS: Prospect Park in Brooklyn, another migration hot spot, was designated an Important

Bird Area in 1998. The 538-acre Forest Park in Queens is also a good place to look for spring migrants. At Jamaica Bay Wildlife Refuge in Queens, more than 330 bird species have been recorded, including numerous spring and fall migrants.

FOR MORE INFORMATION: Central Park Conservancy: http://www.centralparknyc.org.

RECOMMENDED READING: *The New York City Audubon Society Guide to Finding Birds in the Metropolitan Area,* by Marcia T. Fowle, Paul Kerlinger, and William Conway. Ithaca, N.Y.: Cornell University Press, 2001.

SUMMER

5
Summer Splendor

How falls it, oriole, thou hast come to fly
In tropic splendor through our Northern sky?
At some glad moment, was it Nature's choice
To dower a scrap of sunset with a voice?
Or did some orange tulip flaked with black,
In some forgotten garden, ages back,
Yearning towards heaven, until its wish was heard,
Desire unspeakably to be a bird?
— *Edgar Fawcett, "To an Oriole"*

A T DAWN along the shore of Lake Erie, a warbler shakes itself from sleep. It hops along the branches of a willow, stabbing at leaf buds with its bill to pick off tiny insects. When the sun rises, it illuminates the brilliant yellow warbler as he flits to the top of the tree and sings, *Sweet, sweet, I'm so sweet!* For nine years in a row, this yellow warbler with leg band number 1750-17109 has come back to sing near the banding station at the Long Point Bird

Observatory in Ontario, Canada. The staff biologists know him affectionately as Wally. A dynamo weighing less than half an ounce, Wally has traveled back and forth from Canada to Central America each year, about 44,700 miles in his lifetime.

For the past six weeks, Wally flew north from the tropics — perhaps Panama or Costa Rica — crossing more than two thousand miles to get back to Long Point. Now with the journey complete, he will spend most of the next six weeks in a thicket about fifty yards in radius. Within these bounds, he will sing hundreds or thousands of times a day. A pale lemon-yellow warbler, a female, will join him on this territory, gathering hundreds of plant fibers and cobweb strands to build a nest. Later, she and Wally will travel back and forth thousands of times, bringing food to their hungry nestlings. Working with urgency, they will make the most of the brief Canadian summer, their one chance all year to mate and raise young.

Yellow warblers breed all across North America, from the forests of the Pacific Northwest to the Atlantic Ocean, and from the tundra of Alaska to the Mexican Plateau and Cordillera. Yet Wally always comes back to the same place, passing up hundreds of miles of suitable habitat along the way. Why? In the evolutionary scheme, what matters most is not the number of miles that Wally travels, but the number of young that he raises by the end of life's journey. Wally's ancestors probably returned to familiar places each summer to sing, mate, and raise their families. If the strategy works for Wally, too, future generations will inherit his instincts for travel and fidelity to home.

Like Wally, many songbirds return to the same places

where they have nested before. Researchers on Vancouver Island found that more than 70 percent of American robins came back in subsequent years. In Delaware, about 60 percent of male wood thrushes returned to sing their flutelike songs in a suburban woodland, sometimes claiming the same territories as the previous year. About 50 percent of purple martins reoccupied a martin house in Maryland, some even selecting the same compartment. Return rates of 20 to 60 percent are typical for banded songbirds, a number that would probably be even higher if more of them lived to make it back. Wally, the longest-lived yellow warbler on record, is exceptionally lucky. Each year, the odds are about fifty-fifty that an adult yellow warbler will die.

For birds with short lives and few chances to breed, choosing a place with proven success could be the best way to pick a reliable site. Recent studies show that migratory birds assess their nesting success and remember it when deciding where to settle the following year. In Keith County, Nebraska, cliff swallows nest in colonies, affixing their gourd-shaped mud nests to the undersides of bridges. Cliff swallows that raise many young are more likely to return to the same bridge the next year; those whose nests failed are more likely to switch to another colony nearby. Researchers have found that American robins, wood thrushes, brown thrashers, bobolinks, prothonotary warblers, and hooded warblers are also more likely to come back to places where they have succeeded in raising young.

Returning to a known territory has other advantages, too. Birds are especially motivated to protect terrain they have already fought for and won, giving them an edge over newcomers. They may benefit from already knowing the territory as they search for food and seek shelter. Returning birds may also benefit by settling among familiar neighbors. By broadcasting recorded songs, researchers have discovered that hooded warblers arriving on their territories in spring can recall the songs of their neighbors from the previous year. Similar experiments show that yellow warblers, American redstarts, western meadowlarks, and red-winged blackbirds can also recognize and remember the songs of neighbors of their own species. By distinguishing the voices of known rivals from strangers, birds can avoid needless conflicts and devote their energies to chasing out new challengers.

AS THE DAYS LENGTHEN, birdsong everywhere marks the beginning of competition for mates and territories. The vocal contests begin before dawn, first the robins chirping, *cheerily, cheer up, cheer up, cheerily, cheer up,* then the sparrows, vireos, and warblers joining in. The conductor is the sun, light is the baton, and the birds make staggered entrances depending on when each species perceives its cue. By measuring the eye size of fifty-seven songbird species, researchers in the United Kingdom and Portugal have found that birds with the biggest eyes sing first, presumably because they can see the approach of dawn before the others do.

Birds launch into the dawn chorus singing more loudly and energetically than at any other time of day. For the first half hour, Wally freely sings about twelve different songs, directed toward other males. As the day wears on, he settles into a repeated refrain to communicate with potential mates. Though the pace of singing may flag at midday, many songbirds keep up their vocal vigilance from dawn to dusk. When establishing a territory, a dickcissel spends as much as 70 percent of the day singing. The record champion, a red-eyed vireo, sang 22,197 songs in a single day.

Using audio recordings, researchers have demonstrated just how effective song can be. The rapid, twittering melody of a male house wren, recorded and broadcast by researchers near a nest box, can entice females to come investigate and even begin building nests. Broadcast recordings can also convince intruders to stay away. Great tits, European relatives of chickadees, repeat a piercing, singsong *teacher teacher* to keep rivals away. When researchers temporarily removed several great tits from their territories, rivals respected the territorial boundaries, at least for a while, if they could hear recordings of their neighbor's song. In contrast, they readily intruded on territories that had fallen silent.

A songbird without a song is an anomaly, ignored by mates and rivals alike. In 1983 and 1984, researcher Mary McDonald watched as twenty-one seaside sparrows sang silent songs in a Florida salt marsh. They puffed up their chests, threw back their heads, and made all the motions of song, but the air escaped through a tiny, temporary puncture that McDonald had made in a membranous air sac involved in producing sound. Nearby, other sparrows were uttering their full-throated trills, warbles, buzzes, and whistles. Some had been spared McDonald's procedure, and others had been captured and subjected to a mock surgery that left the air sac intact.

Twenty of the silently singing males subsequently lost

their mates, whereas none of the fifty normally singing males did. All of the singing males kept their territories, but all of the songless males lost portions or all of their territories. Fortunately, most of them reconquered their territories when their voices returned. The next year, they came back, sang normally, and attracted mates as if the strange ordeal had never happened.

Using playback experiments to see how birds respond to one another's songs, researchers have also shown that a bird's success depends on the way that it sings. A song sparrow signals that he wants to escalate conflict by matching his song type with the same one sung by his neighbor. He switches to a different type if he wants to back down. The dialogue helps determine whether a bird keeps his territory and even influences how long he lives. A song sparrow sings about nine songs on average, but the repertoire differs from bird to bird. Song sparrows that don't share any songs with their neighbors end up in more fights than those that share two or more songs. These scuffles may take a long-term toll: males with different repertoires have a survival rate that is only about half that of males who sing one another's songs.

Mockingbirds impress mates and rivals with vocal variety; an average male can sing two hundred different songs. They learn new songs throughout their lives by imitating the songs of other birds and sounds they hear from day to day, including squeaky gates, barking dogs, sirens, and car alarms. Mockingbirds with the largest repertoires on the Edwards Plateau in Texas have the best territoires, laden with food such as insects, wild grapes, and persimmons. They also find mates faster than males with more limited vocabularies. Researchers speculate that a male's repertoire showcases his learning ability, longevity, or experience as father and mate.

Song is intrinsic to songbirds; their very anatomy is adapted for it, from their voice boxes to their brains. Specialized control centers in the brain help songbirds produce and remember complex melodies. In the 1980s, researchers made the startling discovery that the song-producing region of a male canary's brain grows and shrinks each year, with new neurons regenerating as needed to learn and execute song. Songbirds also have a specialized syrinx, the sound-producing organ that consists of two voice boxes where the trachea divides in two. With extra syringeal muscles and membranes, songbirds have exquisite control over the sounds they make.

A wood thrush can duet with itself by controlling the two sides of the syrinx independently, uttering both high- and low-pitched strains at the same time. A cardinal whistles an upsweeping note by using its left voice box to produce the lower notes, then seamlessly continuing with its right voice box to produce the higher notes. In a single liquid song, a winter wren utters hundreds of different syllables in a precise and rapid sequence—and delivers it with ten times the power of a crowing rooster, per unit weight. A single virtuoso brown thrasher can sing two thousand different songs.

For birds the goal is simple—to secure a territory, to win a mate, to contribute the only lasting legacy of their brief lives—the passing on of genes to the next generation. When birds sing, they are on a mission, wrote Donald Kroodsma, a professor emeritus at the University of Massachusetts,

Amherst, and a leader in the study of how birds communicate. The mission is to survive and reproduce, each note of sound improving the bird's chances. According to Kroodsma, "Nearly all of what we hear from the birds on their mission is the negotiating, cajoling, persuading, and impressing that accompany the selfish games required for success."[1]

THE SAME COULD BE said of the color of a bird's feathers. Many male songbirds have extravagantly colorful plumage: the yellow warbler in bright yellow with bold reddish streaks on his breast, the Baltimore oriole in fiery orange and black, the indigo bunting enrobed in shimmering blues, the scarlet tanager in red and black. In *The Descent of Man, and Selection in Relation to Sex,* Charles Darwin wrote, "On the whole, birds appear to be the most aesthetic of all animals, excepting of course man, and they have nearly the same taste for the beautiful as we have."[2] More than one hundred years later, researchers have gathered the experimental evidence showing that the aesthetic of female birds — often modestly plumed themselves — has indeed been an important force in the evolution of male songbirds' plumage. When given the choice, females usually choose the brightest, most colorful suitor, or the one with the longest tail and most ornate plumes. These traits can signal a male's quality — his longevity, health, or ability to help raise their young.

For example, female house finches select males with the brightest red plumage rather than suitors with duller shades of orange or yellow. A red badge apparently signals a male's health, vigor, and quality as a provider. The reddest males survive the winter better and have fewer feather mites, pox lesions, and infections. During the breeding season, the reddest males offer more food to mates who are incubating eggs. They also feed their chicks more often than do males of other hues.

Why do bright colors signal a good mate? House finches, northern cardinals, and certain other finches get their hues from the same pigments that make carrots and apricots orange and autumn leaves yellow. The finches ingest these pigments, called carotenoids, when they eat fruits and seeds. When cardinals eat wild grapes, they consume the pigments that will turn their plumage red when it is time to molt and grow new feathers each autumn. Studies have shown that in years when cold spells destroy grape crops, cardinals molt into duller shades of red.

Similarly, experiments with captive birds show that changing a house finch's diet can change the color of its plumage. Auburn University biologist Geoffrey Hill added carotenoids to the water of captive house finches during the fall molt. Their new feathers unfurled with a striking wash of red, even though initially some of the house finches had been orange or yellow. Evidently in the wild, dull males have more difficulty finding food with carotenoids or utilizing the pigments they do ingest. Females looking for the best mate can gauge their suitors based on their plumage. When females prefer brighter males, they leave a double legacy to their offspring. Sons inheriting the genes of their fathers will have colorful plumage, too; daughters may inherit their mothers' liking for colorful mates.

This preference has favored the evolution of brightly colored male songbirds, sometimes strikingly different in appearance from females and young birds of the same species. John James Audubon painted the male black-throated blue warbler with its striking black face mask, dark blue crown and back, and black stripe down its side, contrasting with its snowy white belly. He painted the juveniles (which resemble females) as a completely different species — "pine swamp warbler" — greenish gray on top, whitish underneath, and with a yellowish stripe running over each eye. Audubon also understandably named several species that turned out to be immature or female warblers of species that had already been discovered, including Bonaparte's flycatcher (Canada warbler), and Louisiana warbler, Children's warbler, and Rathbone warbler (all yellow warblers).

PERSUADED BY THE PROMISE of song, brilliant plumes, and a share of the territory, Wally's mate begins building the nest. Searching for a nest site, she sits down in the V-shaped fork of a willow, as if to try it out for size. She flies back and forth, bringing nettles and bits of plant down for the base, weaving the strands around small twigs of the fork to keep it in place. She collects strips of bark to cover the outside, tucking plant down and fibers to the exterior. After binding the nest together with spider silk, she anchors it to supporting branches, then constructs a rim of grass. Finally, she lines the nest with the fluff from dandelion seeds, willows, and cattails, and the hair shed by deer and rabbits. After about four days of work, she completes the nest and lays a grayish egg with brownish spots around the end. Each morning she lays another until there are four or five.

For the next eleven days, she keeps the eggs against the

skin of her belly, warming them to ninety-nine or one hundred degrees. Occasionally she gets up and hurriedly gleans insects from the leaves of nearby shrubs. Wally comes by several times an hour with food in his bill to spare her the trouble of getting up. On the eleventh day, one of the eggs begins to roll peculiarly in the nest. It bulges in the middle and begins to crack; within minutes a tiny nestling emerges, pushing and kicking with its head and feet to free itself from the shell.

Naked except for bits of smoky gray down fluff, and with eyes closed, it can do little more than rest on its bulging belly, lift its head, and open its mouth. That one ability, however, is all it needs to prompt its parents for the food it will need to grow from a helpless quarter-ounce hatchling to a feathered bird that will hop out of the nest and take flight eight to ten days later.

Its parents, meanwhile, devote tremendous effort to fuel the transformation of the entire brood from hatchlings to fledglings. One observer counted a male yellow warbler bringing food 813 times until the nestlings were a week old; his mate visited the nest with food 1,560 times from after hatching until the young birds left the nest. To maximize each trip, parent songbirds usually carry at least several insects at a time. In a single day, a pair of tree swallows feeds their brood as many as 8,000 prey items, mostly flying insects.

At the same time that birds are flitting back and forth to their nests, they must watch for predators—snakes, chipmunks, squirrels, mice, raccoons, jays, crows, shrikes, hawks, owls, and others. They fearlessly fly toward predators many times their own size in an attempt to drive them off, or spiral to the ground to feign injury and lead the predator away from the nest. Cowbirds also take a toll on yellow warblers and other songbirds by leaving their own eggs for other birds to raise. In Ontario, cowbirds laid eggs in 41 percent of yellow warbler nests. If a yellow warbler notices cowbird eggs mixed in among her own, she may abandon the nest and start over elsewhere, or bury the eggs by constructing a new nest floor and laying another clutch of eggs on top. One yellow warbler buried cowbird eggs five times, building a six-tiered nest with a total of eleven cowbird eggs sandwiched in between. Despite their best efforts, songbirds lose many of their eggs and young to cowbirds, predation, starvation, and extremes of weather. One study in Manitoba found that only 35 percent of yellow warbler eggs produced fledglings.

As male and female songbirds work together to raise their young, they seem to epitomize a devoted partnership. Traditionally birds were viewed as a model of monogamy. Before the availability of molecular techniques to analyze genetic relationships, to the best of any ornithologist's knowledge, some 93 percent of taxonomic songbird families were monogamous, with just one male and one female attending each nest. However, genetic studies since the 1980s have turned that estimate upside down. As of 2002, only 14 percent of songbirds surveyed using DNA have proved to be truly monogamous. For example, among nests of reed buntings, an Old World species, 86 percent of broods contained at least one chick not sired by the male attending the nest! On average among "monogamous" birds, 19 percent of broods include at least one offspring who has a different father than its nestmates.

DNA studies and field observations show that when you

account for all of a male's *genetic* offspring, which may be scattered about in different nests, song and plumage matter even more than at first glance. As many as 54 percent of yellow warbler nests, for example, have nestlings sired by a male other than the one bringing food to the young. By matching the genetic profiles of nestlings to males in the area, researchers have found that males with more reddish streaking on their breast plumage are able to cuckold males with less streaking more often than vice versa. Other studies have shown that males with the most colorful plumage or extreme feather ornamentation sire the most offspring by mating with more than one female. In their preferences for mating with brilliantly colored males — even if not the ones with whom they are paired — female songbirds have driven the evolution of their mates' colorful plumes.

If Wally succeeds, soon he will be following around young warblers from his nest — at least some of which are his biological offspring. He and his mate will bring them food until they learn to fend for themselves. In their muted olive-gray and lemon-yellow plumage, the young birds will strike out on their own for the tropics; next spring, they will find their way north to look for a place to raise their own families.

Summer is the crucible, the time of invention. The instinct about where and when to return, the power of song, and the brilliance of a bird's feathers are shaped by whoever fledges the most young and passes on the most copies of their genes to the next generation. After a bleak northern winter, birds arrive in what seems to be tropic splendor, bringing fanciful songs and brilliant colors from exotic lands. In fact, it's the drama here on the breeding grounds — in the bushes

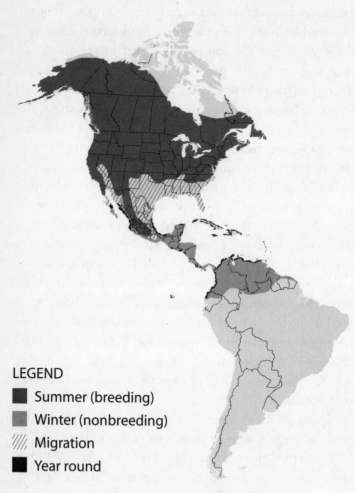

LEGEND
■ Summer (breeding)
■ Winter (nonbreeding)
/// Migration
■ Year round

Yellow warbler migration map.

of our front yards, in the abandoned fields beside our highways, in the parks and wild lands of our temperate landscapes—that ultimately makes birds what they are—musical, colorful, and keen to come back to the places where their own young have flown from the nest and taken to the sky.

6

WARBLERS AND WOODLAND
INTRICACIES

Each species, including ourselves, is a link in many chains . . .
Land, then, is not merely soil; it is a fountain of energy
flowing through a circuit of soils, plants, and animals . . .
When a change occurs in one part of the circuit, many other
parts must adjust themselves to it.
— *Aldo Leopold, from* A Sand County Almanac

IT'S LATE JUNE in the White Mountains of New Hampshire, and the woods are quiet in the noonday heat. Above the thin hum of mosquitoes, the unhurried song of a bird punctuates the stillness. *Cherr-o-wit, cheree, sissy-a-wit, tee-ooo.* "Red-eyed vireo," says Dick Holmes, resting his hand on the trunk of a maple tree and turning his head in the direction of the bird. He pauses. "Listen carefully. You can hear the frass falling—caterpillar droppings hitting the leaves as they fall through the canopy."[1]

Nearly inaudibly, a dry patter showers downward

through the leaves each time the breeze rustles through the trees. The sound of this sprinkling of digested leaves means that thousands of caterpillars are inching around, eating, growing, and sustaining an assemblage of vireos, warblers, and other songbirds in the canopy above.

Through caterpillars and other invertebrate prey, the birds are linked to the substance and rhythms of the earth. Elements from weathered rock are taken up by plants and passed from leaves to caterpillars and from caterpillars to birds. Researchers have traced in the bones of black-throated blue warblers distinctive signatures of strontium from the rocky till of the Northeast and from the marine and limestone rock of the Caribbean, elemental evidence that these tiny songbirds are derived from the soils of far-flung places. As the seasons change, black-throated blue warblers follow the bounty of insects, traveling back and forth from tropics to temperate woodlands, their numbers fueled by the abundance of caterpillars in summer.

Here in the Hubbard Brook Experimental Forest, the woods are a disorderly wilderness, but as Holmes strides down the hill he seems to know exactly where to go. Since 1969, Holmes, a Dartmouth College biologist, has spent summers among these beeches and maples, following the fates of songbirds to answer a deceptively simple question: What causes the numbers of migratory birds to swell or diminish from year to year and through the decades?

The story that Holmes and his colleagues have unraveled is complex; the fortunes of birds are governed by local, regional, and global factors that are constantly in flux. For reasons that are still a mystery, the abundance of caterpillars fluctuates in northeastern forests, glutting the woods with food for birds and their hungry young, followed by years of relative scarcity. Heeding an unexplained rhythm, northeastern deciduous trees yield prolific seed once every four years on average, fueling booming populations of small mammals that prey on bird eggs and nestlings. As far away as Polynesia, the winds over the ocean shift every four to seven years, triggering worldwide cycles of flooding and drought — cycles that determine the scarcity and abundance of insects and the fates of birds. And the varied landscapes where the birds breed, winter, and travel, are continually changing because of human alterations and natural cycles of aging and renewal. Even as conservationists work to protect birds and other wildlife, they have only meager data about how factors such as these are affecting most species, or about the magnitude of habitat destruction and global warming that could ultimately disrupt the balance.

As Holmes walks through the forest, a black-throated blue warbler sings out from the creek bed below: *Zee zee zee zeet . . . Zee zee zee zeet.* Though the warbler is hidden among the trees, Holmes knows from its location that it is probably "G/Y,A," a male with a green band above a yellow band in its left leg and a numbered aluminum band on its right. Holmes points to a hand-drawn map on a piece of grid paper, where an X marks the nest's location. The records show that the nest was abandoned with one hatchling and two eggs. The female — marked with a red band on her left leg and aluminum and white bands on her right — is missing.

"G/Y,A" continues singing as solitarily as though he is the only black-throated blue warbler for miles. The map shows otherwise — some forty pairs of black-throated blue

warblers inhabit the study area, each pair defending a territory of about 2.5 to 10 acres. Almost every pair has a nest
with eggs or nestlings.

Holmes and his collaborators—biologists Nick Rodenhouse from Wellesley College and Scott Sillett from the
Smithsonian Migratory Bird Center—employ eight full-
time field assistants who rise before dawn from March to
July to scour 250 acres of terrain for black-throated blue
warblers. Amid swarms of blackflies and mosquitoes, they
map the birds' territories, capture and band the adults, and
look for nests. Though the nests are usually within reach,
built in shrubs at about knee-height, they are extremely difficult to find. Locating a single nest may take hours of
watching for the elusive female until she disappears into the
shrubs with a strip of bark in her bill. After she has finished
building the nest, she comes and goes like a phantom,
blending into the leaves during the thirteen days when she
incubates her eggs.

Through the dappled light of the forest, Holmes walks
upslope toward a group of trees with orange flagging around
their trunks—a signal that the field crew has found a nest.
Reading notes on the flagging, Holmes says, "Found on
May 20, 0.5 meters above the ground in a viburnum at
marker A.8. Do you see it?" Just off the ground in a viburnum shrub called hobblebush, the nest is barely visible,
neatly built near the central stem and hidden beneath the
leaves. Holmes has to bend down to see into it. It's a tidy
woven basket just the right size for a black-throated blue
warbler and her brood, but it's empty.

Predators have hit more black-throated blue warbler
nests than usual this year. One pair in the southwest corner

of the plot has already built six nests this season, the first five having ended in failure. "The chipmunks and red squirrels are abundant because last fall there was a bumper seed crop," Holmes says. "As the seed crop disappears, they turn to eating eggs and nestlings." About once every two or three years, heavy seedfall from the trees boosts small mammal populations that take their toll on bird nests the following spring.

Predation is the leading cause of nest failure among black-throated blue warblers at Hubbard Brook: out of every ten nests, two to five are raided by predators. Chipmunks and squirrels are not the only culprits. Snapshots from cameras rigged with motion detectors have recorded a diverse set of marauders feasting on quail eggs at artificial nests, including squirrels, chipmunks, raccoons, blue jays, deer mice, flying squirrels, and a black bear.

AFTER INSPECTING the empty nest in the hobblebush, Holmes continues up the slope to join two field assistants who are preparing to band some nestlings. He finds Dai Shizuka and Kara Lathrop sitting on the ground beneath the trees, surrounded by an assortment of data sheets and equipment. Lathrop reaches into a soft cloth bag containing a nestling, then opens her palm. The tiny half-naked nestling looks like an entirely different creature from the elegantly feathered black-throated blue warblers that are its parents. It immediately stretches out its little back end and deposits a dropping, which Lathrop scoops up with a thin metal ruler and plops into a vial with liquid. Preserved in this liquid, the fecal sample may contain some of the answers that

Holmes and others are seeking about the effect of acid rain on songbirds.

In the late 1960s, scientists discovered that rain and snow at Hubbard Brook were ten times more acidic than they should have been. One sample was as acidic as the juice from an orange. The cause was acid rain, formed when water droplets in the atmosphere mix with air pollutants such as sulfur dioxide and nitrogen oxides produced by power plants and motor vehicles. Acid rain harms trees by leaching calcium from leaves and needles, causing the tissues to become more vulnerable to freezing. Acid rain also changes the fundamental chemistry of the soil. It releases toxic metals such as aluminum and leaches away calcium and other minerals needed by plants and wildlife.

Calcium deficiency can stunt the growth of trees, make them more vulnerable to drought, and diminish their resistance to disease, fungi, and insects. Invertebrates, too, need calcium to thrive. The biomass of earthworms may be as much as thirty-four times higher in healthy stands of sugar maples than in stands with acidified, calcium-poor soils. Snails, slugs, millipedes, and sowbugs are also scarce in areas with little calcium, creating problems for birds that derive calcium from these prey. Nestling birds require a lot of calcium for proper growth. Birds also need calcium to lay their eggs; a single clutch contains about the same amount of calcium as the entire skeleton of most small songbirds.

The consequences of acid rain in parts of Europe have been severe. In some regions, entire forests have died after the trees lost their needles and leaves. In areas where open woodlands have replaced forests, birds such as European robins and chaffinches have disappeared. Scientists have also found that in areas damaged by acid rain, resident birds such as great tits, Eurasian nuthatches, and great spotted woodpeckers produce eggs with defective eggshells. Among some species, brittle eggshells cracked before the nestlings could hatch; in other eggs, embryos died from desiccation because the moisture escaped from the abnormally porous shells. Researchers even found eggs held together just by a membrane, the external shell entirely missing. Although the causes are still undetermined, calcium deficiency is suspected to be part of the problem, since eggshells of great tits in affected areas have 15 percent less calcium than normal eggshells.

Researchers now suspect that a characteristic bird of eastern deciduous forests — the wood thrush — could be declining because of acid rain. In 2002, researchers at the Cornell Laboratory of Ornithology found that some of the steepest declines in wood thrush numbers are occurring in areas that typically receive high levels of acid rain and have calcium-poor soils. These changes might be related to changes in forest habitats, nutrient deficiencies of nestlings, problems with eggshells, or other factors.

So far there is little evidence that North American birds are suffering from eggshell deficiencies, as they are in parts of Europe. Could it be that birds are finding snails, millipedes, and other sources of calcium on their wintering grounds and storing it in their skeletons until it can be mobilized for egg laying? In 2000, Joel Blum from the University of Michigan and Hank Taliaferro and Holmes from Dartmouth tested this idea by tracing the origin of skeletal and eggshell calcium of black-throated blue warblers.

To do this, they measured different isotopes of strontium, a trace element that follows calcium in its pathway as it is taken up from soils, incorporated into plants, reconstituted into arthropods, and converted to the bones and eggs of birds. Since different isotopes characterize the soils of the Caribbean and the Northeast, the researchers were able to deduce where the black-throated blue warblers had obtained their calcium. They found that the bones of black-throated blue warblers are built with a mixture of calcium from their wintering and from their breeding grounds, with as much as 60 percent of skeletal calcium derived from the Caribbean. They also found that the bulk of calcium in eggshells must have been consumed at Hubbard Brook or nearby locations.

No one knows how black-throated blue warblers get their calcium. At Hubbard Brook, Holmes has seen female black-throated blue warblers hopping on fallen logs and leaves before the egg-laying period, perhaps in search of the tiny snails that cluster underneath. For now the warblers appear to be finding enough calcium. However, the soil at Hubbard Brook has only half as much calcium as it did several decades ago. Holmes and collaborators warn that birds could be affected if calcium losses continue. Meanwhile, the nestling's fecal sample that Lathrop collected will be analyzed for undigested hard parts of snails and other invertebrates — clues about how the young birds are getting their calcium.

AS LATHROP AND SHIZUKA prepare to collect data on a second nestling, Holmes bids them good-bye and heads back down the slope to check another nest. As he walks, his footsteps kick up a fluttering cloud of small light-brown moths. The fluty notes of a hermit thrush spiral down from the trees and a scarlet tanager sings out its cheery song, *Queer, queery, querit, queer.*

The woods have changed since 1969 when Holmes first began working at Hubbard Brook, and so have the kinds and numbers of birds. A thirty-year study by Holmes and Thomas Sherry showed that songbirds here declined by about 60 percent, from 210 to 220 individuals for every twenty-five acres in the early 1970s, to 70 to 90 by the early 1990s. Of the twenty-four regularly occurring species, half declined, four of them to local extinction. Although songbird declines elsewhere had been blamed on human-caused factors, especially habitat destruction, Holmes and Sherry implicated another powerful force: natural change.

Although the Hubbard Brook forest had been intensively logged between 1905 and 1915, the forest regenerated after the logging stopped. By the late 1960s, American beech trees, sugar maples, and yellow birch towered above the forest floor, forming a dense canopy over low shrubs. In subsequent years, beech bark disease, ice storms, and hurricanes took their toll on the maturing trees. By the 1990s, treefalls and broken crowns created a patchier canopy, allowing taller shrubs and young saplings to grow up underneath, shading out the lowest shrub layers. The four species that disappeared from the study plots — least flycatchers, wood thrushes, American redstarts, and Philadelphia vireos — had thrived in the middle-aged forests. In contrast, three species that prefer older forests became more abundant over the years — ovenbirds, black-throated green warblers, and yellow-rumped warblers.

Meanwhile, the declines in overall songbird numbers

reflected a natural drop in the abundance of caterpillars following an irruption of caterpillars during 1969–71. During
that outbreak, the woods were seething with saddled
prominent caterpillars that defoliated nearly 1.5 million
acres in Maine, Massachusetts, New Hampshire, New
York, Pennsylvania, and Vermont. For every one thousand
leaves, Holmes could find as many as twenty to forty caterpillars. In the years following the outbreak, the same number of leaves yielded only two to four caterpillars. Songbird
numbers were high in years initially following the outbreak, then declined, coincident with the decreasing availability of food. Saddled prominent caterpillars have
irrupted about once every ten years ever since they were first
recorded during the early 1900s until the outbreak in 1970.
As of 2005, there hasn't been another one, but no one knows
why, or when the next irruption might occur.

For thousands of years, the ups and downs of black-
throated blue warblers have probably tracked cycles of caterpillar abundance and scarcity. Even aside from periodic
irruptions, however, the numbers of caterpillars vary from
year to year, with measurable consequences for black-
throated blue warblers. In 1983, for example, caterpillars
were abundant enough that black-throated blue warblers
could raise two broods in a single summer. Each pair produced more than six young on average. The following year,
when caterpillars were scarce, each pair produced only a
single brood of two young on average. Rodenhouse and
Holmes calculated that in every one of four years, black-
throated blue warblers did not produce enough young to
compensate for estimated levels of mortality.

Fluctuations in the food supply are in turn affected by a

cycle of global weather events, the El Niño Southern Oscillation. In the western Pacific, the sun beats down along the equator, warming the waters, and when the trade winds shift every two to seven years, the warm water moves eastward. In the Caribbean where black-throated blue warblers winter and in the North American forests where they breed, El Niño brings drought. Insects and other arthropods become scarce, making it more difficult for birds to find food. Sillett, Sherry, and Holmes showed that, in El Niño years, black-throated blue warblers suffered high mortality rates on their wintering grounds in Jamaica and relatively few young fledged at Hubbard Brook. In contrast, black-throated blue warblers survive better and raise more young during La Niña when the oceanic conditions reverse. These findings have added Neotropical migrant songbirds to a list of animals—including primates, rodents, and arthropods— whose populations or food supplies have been shown to change in response to El Niño.

How well will these populations hold up if global warming contributes to the frequency and severity of El Niño and La Niña? More intense cycles of drought and flooding are especially worrisome for small populations of animals whose numbers might not rebound easily after difficult years. Even for black-throated blue warblers, a widespread and abundant species, the balance among food, weather, changing habitats, and the birds' continued existence is delicate.

To maintain stable numbers of black-throated blue warblers, according to calculations by Holmes and colleagues, each pair of black-throated blue warblers must produce three or four young to offset annual mortality rates of 40 to

50 percent. During 1990–92, black-throated blue warblers inhabiting portions of Hubbard Brook forest with dense shrubs produced 3.6 young on average, enough to keep the population steady or even to increase it. But in areas where shrubs were sparse, they produced only 2.5 young per pair. In years when food or good territories are scarce, more black-throated blue warblers may die than can be replaced. The researchers at Hubbard Brook are investigating how black-throated blue warblers fare at different elevations. Understanding how changes in vegetation, weather conditions, and other factors affect warblers on their breeding grounds may provide clues about how the birds will be affected by climate change in the future.

On his way to check the nest of a black-throated blue warbler, Holmes crosses a thin rocky creek bed, dry now and filled with leaves. He approaches a group of trees marked with orange flagging, the vertical strips pointing toward the hidden nest. Holmes reads the notes on the flagging: "In viburnum, 5 meters ahead in hollow, about 0.3 meters up." He points down toward the shrubs where, if you know exactly where to look, the nest is in clear view. The female, grayish-brown with a yellowish eyeline, is sitting snug on the nest, absolutely still and beautiful. Beneath her, according to the last nest check here, there are two tiny white eggs soon to hatch.

Her presence makes one wonder: When we go out looking for birds in summertime, binoculars pressed to our eyes, how many knee-high black-throated blue warbler nests are we passing without ever seeing them? How many ovenbirds are sitting on nests, hidden on the ground just yards from our shoes? In every layer of this old forest, songbirds are sitting on eggs and young birds are stretching their necks up, begging for food each time their parents alight on the nest.

Every year, Holmes's research team bands some four hundred nestling black-throated blue warblers. Including adult birds, they have kept track of more than three thousand black-throated blue warblers over the years. It's one of the most long-term and holistic studies of any Neotropical migrant songbird, yet it represents only a fraction of the diverse birdlife in this single forest. "We should be conducting long-term intensive studies of ten species," Holmes says. "And we need to study them year-round, not just during summer."

In 1986, after the black-throated blue warblers had flown south, Holmes's team relocated to Jamaica to search for the warblers there. The studies are revealing new insights about the interconnectedness between the two places, essential to understanding how to protect global habitats that the birds need during their annual travels. But the picture will never be complete until the unknown events of migration can be studied as well. "We know too little about what happens in between summer and winter when songbirds scatter across the globe," Holmes says, "It's just too hard to follow them during migration."

The falling frass is a reminder of how the fate of warblers is inseparable from the irruptions of caterpillars, the seedfall of trees, the abundance of small predators, the annual rains, and the seasons that hasten the birds' journeys from north to south and back again. Now, in late June, caterpillars everywhere are eating leaves, and the birds, invisible to us, are growing, enclosed in eggs, or are breathing quietly beneath the warmth of their mothers' feathers. Only at the end of the season, after this deceptive lull, will the woods suddenly seem to come alive with birds everywhere, as the young emerge from their nests and go hopping about, clamoring and quivering for food during the last days before they're independent and heading out beyond the ocean, entering the phase of year during which we know least about what becomes of them.

7
HOW TO FIND AND MONITOR BIRD NESTS

Within a thick and spreading hawthorn bush,
That overhung a mole-hill large and round,
I heard, from morn to morn, a merry Thrush
Sing hymns to sunrise, while I drank the sound
With joy:—and oft, an unintruding guest,
I watched her secret toils, from day to day,
How true she warp'd the moss to form her nest,
And model'd it within the wood and clay.
And by and by, like heath-bells gilt with dew,
There lay her shining eggs, as bright as flowers,
Ink-spotted over shells of green and blue;
And there I witness'd, in the summer hours,
A brood of Nature's minstrels chirp and fly,
Glad as the sunshine and the laughing sky.
     —John Clare, "The Thrush's Nest"

WHEREVER YOU SEE songbirds in summertime, there are probably nests nearby. Watching birds at their nests can give you a glimpse of the remarkable progression from nest building until the fledging of young, typically within a brief span of six to eight weeks.

If you find a bird's nest, whether of backyard robins or wood-dwelling warblers, consider keeping track of the nest and its fate. You can submit this information to a nest record card program or other citizen-science project (see appendix B). Nest records are a valuable resource for ornithologists. Collectively, this information — gathered from thousands of observers from across the continent — can help researchers answer large-scale questions about birds that would otherwise be impossible to investigate.

Scientists have used information submitted to the Cornell Laboratory of Ornithology's Nest Record Card Program to help describe the breeding success and basic biology of birds, including where they build nests, when they lay eggs, and the degree of seasonal, regional, and yearly variation in nesting habits and reproductive success. Thousands of records were summarized in *The Birds of North America,* an eighteen-volume series that presents in-depth information about every bird species that breeds in North America. Researchers have also used nest records to answer specific questions about global warming and other effects of human activities on breeding birds.

Using 21,000 records from the Nest Record Card Program, researchers Peter Dunn and David Winkler found that the breeding cycles of some birds may be affected by global warming. On average, tree swallows during the 1990s laid their eggs nine days earlier than they did three decades earlier. The shifts in lay dates were associated with warmer spring temperatures and suggest that global warming has the potential to affect when birds lay their eggs. In another study, Pennsylvania State University researcher Margaret Brittingham used nest records to help show how cowbirds have a more pervasive effect on songbirds in fragmented forests. The cowbirds were more likely to lay their eggs in other birds' nests in small patches than in large, intact forests.

The Cornell Laboratory of Ornithology's Nest Record Card Program includes more than 300,000 cards and is open for use by researchers. The program no longer has funding to solicit cards from new contributors, but efforts are being made to find new support. People interested in monitoring nests can submit their data to one of numerous other nest record card programs in Canada and the United States, whether they have just one card to submit, or hundreds. If you enjoy finding nests, you may also want to join breeding bird atlas efforts that rely on nest-finding data to inventory a region's breeding birds.

Jim Berry, a personnel worker for the federal government, is one of hundreds of people who have made nest-finding a serious hobby. He looks for nests close to home, as well as on birding trips across the country, and has contributed 856 nest records to science since the 1970s. Berry recalls, among his most memorable experiences, finding the nest of a Say's phoebe underground, four feet down in a well shaft, discovering the nest of a blue grosbeak among shin-high wildflowers in Colorado, and watching American dippers go to nests beneath wooden bridges over streams in the mountainous West. When asked what motivates him to keep looking for nest number 857 and beyond, Berry said simply, "I'm addicted. There's nothing I can think of that I'd rather be doing."[1]

How to Find and Watch Breeding Birds — and Record Your Observations for Science

Every bird species has particular nesting habits, but some general guidelines can help you find nests and ensure the well-being of the birds you watch. Citizen-science projects vary in their specific guidelines for gathering and recording data, so if you intend to submit your observations, be sure to follow the accompanying instructions for that project.

If you enjoy watching birds at their nests, you may also wish to consider providing nest boxes for species that breed in cavities, such as bluebirds, chickadees, titmice, nuthatches, tree swallows, and house wrens. The Birdhouse Network, a citizen-science project of the Cornell Laboratory of Ornithology, offers information about cavity-nesting species, nest box building plans, and instructions for monitoring birds in nest boxes (see appendix C).

Background Check

Learn as much as you can about the species you are looking for, especially its nesting habits. For example, American goldfinches in the East are among the latest breeding songbirds in the temperate zone. It can be frustrating looking for nests in May unless you realize that most individuals don't start nesting until June and July.

By reading about the birds in your area, you may also learn valuable clues about where they usually build their nests and what their nests look like. A brown creeper that seems to vanish into a tree may be sitting in its hammock-shaped nest behind loose flaps of bark, often on dead or dying trees. A meadowlark constructs its nest on the ground in tall grass, often with runways leading up to the nest. Sometimes it fashions an entrance tunnel that leads up to a nest with a woven grassy roof. A barn swallow attaches a nest of mud to a building, culvert, or other structure, but a tree swallow usually settles in a cavity, such as a nest box or a tree excavated by a woodpecker. Background reading can also provide important details to help interpret what you see. For example, a male marsh wren in eastern Washington state builds an average of twenty-two nests in a single season, only one of which contains the eggs.

Learning the Territory

If the species you are looking for is territorial, finding the breeding territory in early spring can help you locate a nest later. Males usually sing the most while they are establishing territories and attracting mates. You can learn the boundaries of a songbird's territory by paying attention to where the male sings. Later you will be able to focus your observations on the territory when you look for the nest.

Construction Zone: Finding Nests in Progress

If you suspect that birds are building nests, find out whether the male, female, or both members of the pair contribute to

Baltimore orioles often inhabit open areas with scattered trees, whether nesting among cottonwoods in summer or sipping nectar from tropical flowers in winter. (Male at left, female at right.)

Rose-breasted grosbeaks breed in Canada and the United States and winter from Mexico to South America. (Female above, male below.)

In summer, **magnolia warblers**
inhabit Canada and the northern
United States, nesting in coniferous
trees such as spruce and hemlock.
They winter in Mexico, Central
America, and the Caribbean. (Female
at left, male at right.)

Chestnut-sided warblers split
their time each year between the
forests of North America and
the forests of Central America.
(Male above, female below.)

Gray-cheeked thrushes breed in the taiga and tundra of Alaska and Canada, and winter in South America. Because of their skulking habits, they are seen infrequently during migration.

In fall, **wood thrushes** migrate to Mexico and Central America.

Yellow warblers migrate thousands of miles each year. Studies of individually banded birds show that they sometimes return to the same territories year after year on both ends of their journeys.

The number of **black-throated blue warblers** fluctuates in cycles due to distant and local weather events, caterpillar outbreaks, seedfall in forests, and nest predators. (Female above, male below.)

By the time a **bobolink** is two and a half years old, it will have migrated a distance greater than if it had traveled around the entire world at the equator. Its brain—about the size of an almond—contains everything it needs to know in order to navigate using magnetic fields, polarized light patterns, and stars. (Male in spring plumage above, male in winter plumage below.)

Scarlet tanagers nest in eastern deciduous woodlands in summer and spend the winter in tropical forests from Panama to South America. Adult males are brilliant red with black wings in summer, but yellowish green with black wings in winter. (Female above, male below.)

Indigo buntings breed in eastern North America. They winter in Florida, Mexico, Central America, and the Caribbean islands, roaming in flocks of hundreds or thousands. (Female above, male below.)

On their wintering grounds in Jamaica, mature male **American redstarts** inhabit mangrove forests, forcing most females and young birds into dry, scrubby areas. (Male at top, female in center, immature male at bottom.)

Black-and-white warblers can be found as far south as Venezuela in winter and as far north as Canada in summer. They inch along tree trunks, stabbing at the bark with their bills. (Male above, female below.)

As long as food is plentiful, **common redpolls** can survive some of the coldest winters in the world. When boreal seed crops fail, usually every other year, redpolls migrate southward from Canada into the United States. (Female at left, male at right.)

Yellow-rumped warblers can winter as far north as Nova Scotia, subsisting on wax myrtle berries and bayberries, a source of energy that other warblers can't digest. In another dietary innovation, yellow-rumped warblers wintering in Mexico drink droplets of nectar from the tip of a hair attached to the back end of an insect that lives on trees. (Female above, male below.)

nest construction. Females of many songbird species build the nest alone, or with some help from the male. If that's the case, focus most of your attention on the female.

If you glimpse a bird with a twig, feather, plant fiber, or other bit of nesting material in its bill, keep watching. Many songbirds make hundreds or thousands of trips back and forth over the course of several days, as they gather materials for their nests. When a bird disappears repeatedly into the same foliage with nest material, it is going to its nest. Try watching with binoculars or a spotting scope from a distance to observe the bird as it constructs the nest. If the bird is hidden from view, you can walk up and scan the vegetation closely for a nest. To minimize disturbance, limit your inspection to a few minutes. If you don't see anything, try again in a day or two, when the nest may be bigger. Beware that some birds, such as ovenbirds, nest on the ground. If you notice them in the area, watch where you step.

Disappearing Acts: Finding Birds with Eggs

Finding an open-cup nest with eggs is more challenging than finding a nest under construction. An incubating bird is quiet and motionless, and it comes and goes from the nest stealthily, sometimes even taking indirect routes to the nest through foliage. Some species of birds are known as "tight sitters." Rather than quietly leaving a nest when a predator approaches, the bird sits motionless and flies off the nest at the last moment (some species, such as cedar waxwings, may remain motionless even when touched). If you suspect that you have flushed a bird off its nest, or see it disappear repeatedly into the same vegetation, scan the area quickly. If you don't see a nest, retreat and watch from a distance. You may be able to use binoculars or a spotting scope to see a bird return to its nest. If the bird does not return within several minutes, leave the area and come back another time.

Develop a "Search Image"

Learn where different species prefer to nest. Do they breed on the ground, in shrubs, or in the canopy? Which plants and portions of plants do they prefer? What do the nests look like? Once you develop a "search image" for a particular species—an expectation of what the nest looks like and where you are likely to find it—you will be able to find other such nests more quickly.

Finding Nests with Young

One of the easiest ways to find a nest is to see a bird with food in its bill and watch where it goes. Most small songbirds feed their young four to twelve times per hour. Even if you are too far away to see food in the bird's bill, you can deduce that it is carrying food to a nest if it repeatedly flies in a certain direction after foraging. Listen carefully, too—you may be able to hear the cheeping sounds of young when they beg for food.

Mob Scene

Are the birds acting agitated, uttering high-pitched alarm calls and fluttering around you or another potential predator such as a hawk, owl, or cat? Mobbing behaviors can reveal in a flash which birds and how many are breeding in an area. The more tenacious they are, the more likely they are to be defending nearby young. Studies have shown that birds are the least aggressive when they have no eggs or young in the area. They become more bold around predators as the nestlings get older.

Signs of Independence

Even after most young songbirds leave the nest, they follow their parents around and beg for food by quivering their wings and calling. If you know or suspect that you have young birds in your yard, you may be able to entice them to come close to your house by putting out food such as suet (peanut butter mixes or commercial formulas that don't spoil in hot weather) or mealworms, and leaving water for bathing and drinking. Make sure that these areas are safe from cats. Because juvenile birds are still learning to fly, they are easy targets for predators.

You've Found the Nest! Now What?

Once you have found a nest, try to see what is in it, but keep disturbance to a minimum. If the nest is too high, you can use a mirror on a pole to see into it. Attach a cosmetic mirror or bicycle mirror to a pole and hold the mirror above the nest. You will probably be able to see any eggs or nestlings in the reflection.

You can also learn a lot about the contents of the nest by observing the parents' behavior, watching from a distance through a spotting scope or binoculars. If a bird sits down at the nest as soon as it arrives, it is probably incubating eggs. If it perches on the edge of the nest and bends down, it is feeding nestlings. If the birds suddenly stop coming to the nest before the young have fledged, the nest has failed.

How often should you look into a nest? Perhaps once is all you need to satisfy your curiosity. However, if you want to check in regularly, you may be able to construct a record that allows you to calculate approximately when the eggs were laid and when the nestlings hatched, even if you didn't observe the nest on the exact date when these events occurred.

Most songbirds lay one egg each day, in the morning. Ideally, find the nest while it is being built. If the nest lacks eggs, visit every other day until you see at least one egg. During the incubation period, visit every three days, an interval that minimizes disturbance but also allows you to narrow down when the clutch was completed, when the eggs hatched, and when predation occurred, if any. Make a visit soon after hatching to count the young, and one more visit near the end of the nestling period to see whether they have survived to near fledging age. If possible, check with binoculars from a distance to see whether the parents are still coming to the nest to feed the young.

Record all of your observations in a log or on a project nest card. Include the date, written descriptions of behaviors, and records of the number of eggs and nestlings in the nest, if known. If a nest fails, look for clues about what might have been the cause. For example, a hole in the bottom could indicate that it was raided by a snake. After the nest is no longer active, take some time to describe where the nest was and any other details about the surrounding vegetation. Keep your eyes open for further signs of breeding activity. Some birds nest two or more times in a summer. Finally, submit your data to a citizen-science project.

Laws and Ethics: Observing Nests Responsibly

For both ethical and scientific reasons, it is undesirable to interrupt the breeding activities of birds. It is also federally unlawful to harm or harass native birds or to possess their eggs, nests, or feathers. This includes eggs and nests that have been abandoned for the season, as well as feathers found in the nest or on the ground. Because it is impossible for anyone else to tell how these items were collected, the law prohibits collecting them at any time, for the birds' protection.

If you have found a species that is endangered or threatened in your area, you may want to inform local coordinators of breeding bird atlases or nest record programs, but do not attempt to find the nests of these sensitive species on your own.

When searching for or monitoring other nests, never break branches or alter the vegetation. Doing so could expose the nest, making it easier for predators to find it. Avoid checking nests during extremes of weather such as heat waves, cold spells, and storms. In good weather, check nests during midday, when predator activity tends to be lower than at other times of the day. Before approaching a nest, watch carefully from a distance. If you see any predators in the area, such as hawks, jays, or cats, avoid going to the nest.

The best moment to approach a nest is after you have

seen the parent leave to forage. Keep your time near the nest to a minimum, never more than a few minutes at once. The birds perceive you as a threat, and your presence is stressful to them. If the birds become agitated by fluttering around and making loud calls, your presence could also attract the attention of nest predators.

Do your best to avoid leaving scent trails that could give away the nest's location to mammalian predators. If you need to part the vegetation, use a stick rather than your hand. Don't leave a dead-end trail at the nest. Loop around or walk past it, then back again so your scent trail leads away from the nest.

All of these precautions are important to minimize disturbance to the birds. That said, many bird species are tenacious defenders of their eggs and nestlings so your fleeting presence is unlikely to cause them to abandon the nest. For the most part, you will be able to walk up to the nest and have a quick look inside without causing any more disturbance than would a crow, jay, or owl if it passed by. What you learn may help you gain a deeper appreciation of birds and a fuller scientific understanding of their lives.

Hot Spots for Summer Songbird Diversity

Summer bird watching is especially exciting where different habitats converge, bringing diverse bird species together in the same region. The Southern Sierra Nevada and Kern River Valley in California, and the deserts, riparian habitats, and mountains of southeastern Arizona are examples, but there are numerous places that span ecological transition zones, including Riding Mountain National Park in Manitoba, Prince Albert National Park in Saskatchewan, and the region including the Pawnee National Grasslands and Rocky Mountains National Park.

Southern Sierra Nevada and Kern River Valley, California

California's Kern River Valley and the surrounding areas offer spectacular contrasts in landscapes and wildlife, from lush riparian woodlands to the towering sequoias of the Sierra Nevada to the Mojave Desert and foothill chaparral habitats. Natural areas in the Kern River region include the Giant Sequoia National Monument, Audubon California's Kern River Preserve, eight federally designated wilderness areas, the South Fork Kern River Globally Important Bird Area, and the Butterbredt Spring Nationally Important Bird Area. The bird checklist for the Kern River Valley and Southern Sierra Nevada shows 342 bird species, including 184 known to breed there.

Within fifty miles, the elevation in Kern County ranges from 206 feet to 8,824 feet above sea level, with corresponding changes in plants and wildlife. On the east side of the mountains, gray flycatchers, plumbeous vireos, Brewer's sparrows, and vesper sparrows inhabit the Great Basin Desert with its sagebrush scrub and pinyon and juniper trees. In the adjacent Mojave Desert to the south, Joshua trees, juniper, cacti, creosote bush, and other high desert plants provide nesting habitat for Costa's hummingbirds, cactus wrens, Le Conte's thrashers, Scott's orioles, and black-throated sparrows. In the chaparral, wrentits, California thrashers, and black-chinned sparrows breed. The forested southern Sierras are home to white-headed wood-peckers, Cassin's vireos, hermit warblers, and MacGillivray's warblers. The dry east face of the mountains are home to pygmy nuthatches, whereas red-breasted nuthatches are predominant on the wetter west side. At the base of the mountains, the Great Valley Grassland includes deciduous woodlands, valleys, hills, and riparian areas, home to nesting yellow-billed cuckoos, summer tanagers, blue grosbeaks, indigo buntings, and yellow warblers.

Audubon California's Kern River Preserve includes riparian habitat with trails open to the public. In a quarter-mile area around the parking lot at the headquarters, visitors can see numerous breeding birds, including oak titmice, bushtits, Bewick's wrens, house wrens, western bluebirds, American robins, yellow warblers, common yellowthroats, summer tanagers, savannah sparrows, song sparrows, blue grosbeaks, lazuli buntings, Bullock's orioles, lesser goldfinches, and Lawrence's goldfinches.

During spring migration, nearby Butterbredt Spring offers a rare glimpse of western species on the move from late April until mid-May. The narrow canyon funnels birds toward the spring as they migrate northward. The best viewing is during the first few hours of daylight, when as many as eleven warbler species can be seen in a single morning. Birders standing on top of the canyon on a good day may witness thousands of migrants flying overhead, at eye level, or in the trees and spring below. On May 19, 1999, an observer recorded 3,000 warblers, including 1,700 Wilson's warblers, 700 yellow warbers, 400 Townsend's warblers, 80 MacGillivray's warblers, and 80 orange-crowned warblers.

As the birds travel northwest, they enter a fourteen-mile stretch of Fremont cottonwood and red willow riparian forest along the South Fork Kern River, a major migratory stopover site. A good place to watch for migrants is the Kern River Preserve's Migrant Corner Trail, which runs along the north edge of the South Fork Kern River riparian forest.

WHEN TO GO: April through May for desert species, June through July for breeding birds.

NEAREST BIG CITIES: The Kern River Preserve is about fifty-five miles northeast of Bakersfield (approximately one and a half hours). Driving time from the Los Angeles International Airport is three to four hours.

SPECIAL ACTIVITIES: Workshops, nature walks, and bird-banding demonstrations are available through Friends of the Kern River Preserve. The annual Kern River Valley Bioregions Festival in late April celebrates the natural and recreational diversity of the area, including plants, wildlife, geology, astronomy, hiking, rafting, fishing, art, and, of course, birding.

FOR MORE INFORMATION: http://www.audubon.org/local/sanctuary/kernriver/about.htm.

RECOMMENDED READING: "Valley Wild: Tips for Birding the Southern Sierra Nevada and Kern River Valley and Annotated Checklist," by Bob Barnes, 2005, http://www.valleywild.org/krvbirdtips.htm.

The San Pedro Riparian National Conservation Area and the Huachuca Mountains, Arizona

Southeastern Arizona is an extraordinary region for birds, with 514 species recorded, 400 of them regularly — all together more species than in any other interior area of similar size in the United States. The region includes portions of the Rocky Mountains, Chihuahuan Desert, Sonoran Desert, and Sierra Madre Mountains, including desert scrub and grassland, oak woodland, pine forest, and riparian forests.

The San Pedro Riparian National Conservation Area is a Globally Important Bird Area along the San Pedro River, a magnificent cottonwood and willow corridor through which an estimated 4 to 6 million birds migrate every year. Some 250 bird species have been found along the upper San Pedro River, including more than 100 breeding species, such as vermilion flycatchers, summer tanagers, yellow warblers, common yellowthroats, yellow-breasted chats, and blue grosbeaks. Adjacent desert grasslands have nesting canyon towhees, Abert's towhees, and Cassin's, Botteri's, and black-throated sparrows. To the southwest, the grasslands meet the Huachuca Mountains with canyons that ascend from the oak savanna and Madrean oak woodland to pine-oak forest. Brown, Carr, Garden, and Miller canyons and the Ramsey Canyon Preserve have trails that are open to the public. Acorn woodpeckers, Arizona woodpeckers, bridled titmice, Virginia's warblers, and black-throated gray warblers live in the oak woodlands of the foothills. Painted redstarts and Scott's orioles can be seen among Arizona sycamore trees and bigtooth maples bordering riparian corridors.

At higher elevations among pines and junipers there are olive warblers, Grace's warblers, red-faced warblers, western tanagers, and red crossbills. The canyons are also famous for their stunning array of hummingbirds. Fifteen species of hummingbirds and various hybrids visit the feeders at Beatty's Miller Canyon Guest Ranch and Orchard in Miller Canyon. Rarities such as eared quetzal, Aztec thrush, and tropical parula have been spotted along the trail in the upper part of Miller Canyon. These are just a few of the places in southeast Arizona that capture a spectacular diversity of birds within a small area. The Southeastern Arizona Bird Observatory has a list of fifty birding locations on the Southeastern Arizona Birding Trail, along with descriptions and directions, at http://www.sabo.org.

WHEN TO GO: Mid-April to mid-May for spring migration. Mid-April through early July for nesting species. Some species nest for a second time or delay breeding until the rainy season, mid-July to September. Mid-September is the best time for watching mixed flocks of songbirds during fall migration.

NEAREST BIG CITY: Tucson International Airport is about seventy-eight miles from the San Pedro National Conservation Area and about eighty-one miles from the Ramsey Canyon Preserve.

SPECIAL ACTIVITIES: The Southeastern Arizona Bird Observatory offers guided bird walks, tours, and educational workshops at various locations, including Miller Canyon and the San Pedro River. Personalized guide services are also available. The observatory sponsors the annual Fiesta de las Aves (Festival of the Birds) in late April and early May. The festival, a celebration of migration, includes field trips at various locations. In the fall, the regional Southwest Wings birding festival offers workshops, field trips, and other programs.

FOR MORE INFORMATION: Southeastern Arizona Bird Observatory: http://www.sabo.org.

RECOMMENDED READING: *A Birder's Guide to Southeastern Arizona,* by Richard C. Taylor. Colorado Springs: American Birding Association, 1999.

AUTUMN

8

*F*IVE THOUSAND MILES SOUTH BY NIGHT

Gayest songster of the spring!
Thy melodies before me bring
Visions of some dream-built land,
Where, by constant zephyrs fanned,
I might walk the livelong day,
Embosomed in perpetual May.
Nor care nor fear thy bosom knows;
For thee a tempest never blows;
But when our Northern summer's o'er,
By Delaware's or Schuylkill's shore
The wild-rice lifts its airy head,
And royal feasts for thee are spread.
And when the winter threatens there,
Thy tireless wings yet own no fear,
But bear thee to more Southern coasts,
Far beyond the reach of frosts.
—Thomas Hill, from "The Bobolink"

WHEN ORNITHOLOGIST Olin Sewall Pettingill was a boy in Maine, his mother led him into a field where she had found the nest of a bobolink. They walked carefully through the grass until she stopped, bent down, and pressed aside a clump of clover to reveal the first nest of a wild bird he had ever seen. It made a lasting impression — along with his early memories of June mornings when the male bobolinks, in their striking black and white plumage, rose into the air and sang in flight above the grassy hillsides sprinkled with red clover and white daisies.

Later, Pettingill would study bobolinks in the fields and meadows near the University of Michigan Biological Station. In May, he watched as the males arrived to claim a patch of meadow, circling above on quivering wings and singing their rollicking, tinkling medleys. When the yellowish-brown females arrived a few days later, the males chased them about the meadow until the females, consenting, dropped down into the grass to mate. In June, Pettingill looked for the nests, shallow cups of grass in a rut, hoof print, or other shallow depression. He counted the eggs — usually four or five — and noted how, just nine days after hatching, the young bobolinks crawled out of the nest and vanished in the grass. Days later, he watched as the young birds flapped their wings to rise above the grass for their first view of the world. In July, the parents followed their young around the open fields, feeding them summer's last grasshoppers and caterpillars. Soon after, the bobolinks all gathered up into flocks and disappeared.

In nearby marshes, Pettingill found them silent and hidden among sedges and cattails. With the shortening days

signaling the coming of fall, the birds' hormones were beginning to change, orchestrating the transformation that
would prepare them for a journey of more than five thousand miles to South America. In the marsh, they lost their
feathers and skulked about with their new plumage only
partly grown in. When they were done molting, they would
be clothed in more than a thousand new feathers. Brown
and olive-buff, the males would resemble the females and juveniles, their bold plumage no longer needed for courtship.
As fall advanced, the bobolinks would grow more restless,
until by night they would fly up from the marshes and depart, uttering their metallic *clink* notes in the darkness.

From all across the fields and grasslands of southern
Canada and the United States, bobolinks embark on one of the
longest migrations of any North American songbird. In fall,
they travel along two main routes to South America. In August
and September, they flood into the southeastern United States,
then down to Florida and across the Caribbean. Others travel
east until they reach the coast from Nova Scotia to Virginia,
then launch out over the Atlantic, riding strong tailwinds. Far
out at sea, six hundred miles from the nearest land, records of
bobolinks flying high over the Bermuda Islands indicate that
these birds make a nonstop crossing to the coast of South
America, a route that is also taken by another phenomenal
migrant, the blackpoll warbler. After making landfall, the
bobolinks continue onward, crossing the equator and reaching the broad grasslands and marshes of southwestern Brazil
and Paraguay by November. As the austral summer advances,
they follow the ripening grain southward, arriving in Argentina and Uruguay in January.

Like all Neotropical migratory songbirds, bobolinks

embark on their first journey when they are less than a few months old. One nine-year-old bobolink was estimated to have flown a distance equivalent to four and a half times around the earth. At the end of each round-trip, the bobolinks return with startling precision: as many as 70 percent of banded males return to the same area where they nested before.

How do bobolinks find their way and back across those immense distances? It's a daunting feat even by human standards. It would be ludicrous to ask a human teenager to go from a patch of meadow in Maine to an Argentinian marsh and back with no instructions, map, compass, or transportation. The bobolink, though, is far better equipped for the trip. In its genes, it carries all the instructions it needs. Weighing less than two ounces, its body includes a built-in magnetic compass, as well as the fuel and mechanics to power its journey. Its brain—about the size of an almond—contains everything it needs to know to navigate to its destination using magnetic fields, polarized light patterns, and stars.

BEFORE MIGRATING IN FALL, songbirds undergo physiological changes that prepare them for migration. As the autumn days shorten, photoreceptors in a portion of the brain called the ventromedial hypothalamus respond to the changing day length, triggering hormonal shifts that cause birds to molt their feathers, eat with fervor, and become restless. Most migrants grow fresh plumage before their long-distance flights. They also abandon their summer diet of insects and begin gorging on fruit or grain, which is more easily converted to lightweight, energy-rich fat. Bobolinks consume 38 percent more food in autumn than they do in summer and switch from eating 90 percent insects to 90 percent grain. The fat accumulates all over their bodies — mostly between the skin and muscle of the abdomen and throat, but also in organs such as the heart and liver. By the time they reach Florida, the bobolinks will have gained about half of their original body weight, with enough fatty fuel to cross the Caribbean.

As they travel, flocks of hungry bobolinks stop over in marshes, grasslands, and fields to devour grain. During the early 1900s, flocks of tens of thousands of bobolinks descended on southern farmers' rice crops. Known as "ricebirds," they usually arrived in late August during the "milk" stage, after the rice had flowered but before the grain had hardened. According to one observer, the birds ate by day and moonlit nights and became so fat and lazy they could hardly be dislodged from the rice.

Many farmers dislodged the birds by shooting into the flocks from dawn until sunset during the entire month when bobolinks occupied the fields. At night, hunters rode in skiffs along irrigation ditches, plucking the birds from the plants where they roosted and bagging them by the dozens. About 720,000 bobolinks were shipped as game from Georgetown, South Carolina, in one year, according to the chief game warden in 1912. Six years later, the Migratory Bird Treaty Act outlawed killing migratory birds in the United States and Canada, but bobolinks continued to be shot as pests, sold as food, and trapped as caged birds elsewhere. In Jamaica, the bobolink was a delicacy known as the "butter bird." "It is but a mouthful, but a luscious and delightful one," wrote one observer during the 1800s.[1]

No matter the dangers that await, bobolinks and other long-distance migrants have only one option—to fly in autumn, the beating of their wings and the path of their flight determined by genetic mandate. For centuries, owners of caged songbirds have known that captive birds become restless in spring and fall, just as wild birds do. As songbirds take to the sky just after sunset, the captives begin flitting against their cages, too. Overtaken by a powerful urge, they keep flapping their wings until early morning when most wild birds land. Their restlessness continues every night, finally ceasing at about the time when the wild birds finally reach their wintering grounds. Ornithologists know this migratory restlessness by the German name, *Zugunruhe*. Like molt and weight gain in fall, *Zugunruhe* is triggered by changing day length, ensuring that the birds are ready to leave at the same time every year.

Scientists also noticed that the captive birds oriented in the same direction they would have taken if they had actually been migrating. Researchers Eberhard Gwinner and Wolfgang Wiltschko in Germany found that hand-raised European garden warblers flew toward the southwest side of their cages while freely migrating birds were traveling southwest to Spain and Portugal. When the migrating garden warblers later switched headings to turn south into Africa, the captive garden warblers flew toward the south side of their cages. These and similar results with other birds have shown that some migratory songbirds have an innate urge to fly in a specific direction for a set distance—the inherited set of instructions that leads them to their wintering grounds. This discovery explained how young birds find their way on their first migration.

It also explained why some birds take indirect routes to their wintering destinations. As the breeding range of bobolinks expanded over time, birds kept retracing the migratory routes of their ancestors. As a result, bobolinks that breed as far west as North Dakota are believed to migrate southeast toward the Atlantic, rather than heading on a straight line to Brazil. Some bobolinks, however, travel from the westernmost reaches of their breeding range and across the Pacific to South America. As early as 1835, Charles Darwin recorded bobolinks on the Galápagos Islands, six hundred miles west of Ecuador. Additional records since then suggest that these western bobolinks migrate over the Pacific on a path that probably leads them to Peru and Chile.

Many songbird species with widespread distributions follow different migratory routes depending on their origin, and genetic experiments show that these routes are handed down from generation to generation. If an Old World warbler, the blackcap, originates in Germany, it will migrate to the southwest and spend the winter in the western Mediterranean. However, blackcaps from Austria fly to Africa, a route that requires them to fly southeast, then change headings to the south. To test whether these differences were heritable, researchers Peter Berthold and Andreas Helbig interbred German and Austrian blackcaps. In fall, the young blackcaps tried to fly in a direction in between the headings of their parents and switched direction partway through, as their Austrian parent did. In the wild, blackcaps from Germany and Austria don't interbreed. If they did, their young would follow mixed messages, possibly ending up in a disastrous flight over the Sahara desert.

Although songbirds rely on an inborn message when migrating for the first time, they eventually learn to navigate so that they can get back to specific places even when blown off course. During the 1950s, ornithologist A. C. Perdeck captured and banded eleven thousand starlings in Holland, then packaged them up in bamboo cages and sent them to Switzerland by airplane. After releasing the birds in Switzerland, he awaited word from hunters, ornithologists, and others about where the birds were being seen. He reconstructed the routes the birds had taken, based on 354 reports of banded starlings. Young birds kept migrating in the same direction that they had been heading before they were relocated and ended up far from where they should have been. This meant that they must use a compass to travel along a specific heading, but had no map with which to navigate to a particular place. Adult starlings, however, found their way to the wintering grounds despite the detour. Perdeck concluded that these older birds were able to use previous experiences to correct their trajectories.

Bobolinks also have a remarkable ability to navigate, as William Hamilton III learned by accident in 1960. The previous fall, Hamilton had asked a friend in Kenmare, North Dakota, to capture a bobolink and airmail it to him at the University of California, Berkeley, more than a thousand miles to the southwest. Hamilton wanted to test whether bobolinks could use the stars to correct their orientation after being displaced. He planned to observe the bird's heading in a cage with a Plexiglas lid that allowed her a full view of the sky from the rooftop of the Life Sciences building. Before he could complete the experiment, however, the bobolink escaped from her cage. The next spring she showed up in the same North Dakota field where she had been captured before she was shipped to California.

Experiments during the 1950s and 1960s showed that birds could use a sun compass to navigate by day and stars to find their way at night. During the 1950s, scientist Gustav Kramer showed that caged European starlings oriented in a particular direction in relation to the sun and switched directions when he changed the apparent position of the sun using mirrors. In 1957, Franz and Eleanor Sauer conducted experiments in a planetarium that showed European garden warblers could orient to the night sky. When the planetarium simulated the natural sky, the caged garden warblers fluttered away from the North Star, as if headed south toward their wintering grounds. When the Sauers recast the stars so that the North Star appeared in the south, the warblers reversed direction. Later experiments by Stephen Emlen during the 1960s showed that birds did not cue in to the North Star itself, but rather to the stars that rotated around it. When Emlen raised young indigo buntings beneath a sky with stars rotating around Betelgeuse, the buntings oriented away from Betelgeuse just as if it had been the North Star.

To navigate by stars, birds must have a clear view of the sky, but they often migrate beneath clouds. As early as the 1850s, European scientists had suggested that some animals might be able to sense the earth's magnetic field, but it wasn't until the 1970s that researchers produced convincing evidence that homing pigeons had this ability. In the 1960s, Friedrich Merkel and Wolfgang Wiltschko observed that European robins hopped in the same direction that they would have flown on migration, even in a room with no clues from the sun or stars. Using Helmholtz coils,

Merkel and Wiltschko changed the direction of the magnetic field and found that the birds switched directions, too. Since then, at least eighteen species of migratory birds, including bobolinks, have been shown to orient using magnetic fields, an ability that is now believed to be widespread among migratory songbirds. Other animals—including bees, spiny lobsters, sharks, salamanders, turtles, and whales—also respond to changes in magnetic fields.

In 1984, researchers Robert Beason and Joan Nichols discovered that nasal tissues of bobolinks contain magnetite, a magnetic mineral that can act as miniature compass needles. Magnetite has also been found in other animals that respond to magnetic fields, including white-crowned sparrows, bees, and some bacteria, but exactly how animals use magnetic fields to navigate is still a mystery.

Experiments by Beason and colleague Peter Semm have shown that changes in the magnetite could be detected by trigeminal nerves involved in relaying messages from the bobolink's facial areas to its brain. These nerves respond to changes in the magnetic field, though it's unclear exactly how the magnetite would stimulate the nerves. Beason also established that in autumn, caged bobolinks oriented to the south-southeast, but when the magnetic field was reversed, the bobolinks tried to head north. By remagnetizing the birds, Beason and his students caused the birds to orient in different directions depending on the orientation of the applied magnetic pulse. When Beason and Semm used Novocain to numb the nerve they believed was relaying information from magnetite particles to the brain of magnetized bobolinks, the birds changed direction and headed southeast again. This meant that the nerve had allowed them to sense the change in the magnetic field. However, because the bobolinks were able to reorient to their original southeasterly direction even without cues from the sun or stars, Beason and Semm concluded that the bobolinks could probably sense the magnetic field in a second way.

Recent evidence suggests that songbirds may actually see magnetic fields. In 1987, Beason and Semm discovered that bobolinks have neurons in a part of their brain involving vision that respond to a changing magnetic field. In complete darkness, the neurons don't respond, so they need light to function. Not just any light will do, either — it has to fall within a particular range of wavelengths. In yellow-orange light, bobolinks can orient in a magnetic field, but they become disoriented in red light. Wolfgang Wiltschko and colleagues found that Australian silvereyes and European robins also become disoriented in red light, but have a normal sense of direction in blue, green, and white light.

In 2004, a team led by Henrik Mouritsen identified light-sensitive pigments in the retinas of European garden warblers that influence how sensory cells interact with the magnetic field. Because of the curvature of the retina and the variation in the position of the sensory cells, various parts of the retina would respond to the magnetic field differently, allowing birds to perceive the magnetic field as visual patterns. If the birds perceived the magnetic field as blue, for example, they might see a bluish tinge when facing north or south, but not when facing east or west. However, no one really knows what a magnetic field might look like or even whether birds do in fact see it. Although the visual system is involved in sensing magnetism, researchers are not sure whether birds perceive the magnetic field with their vision or through some other unexplained sensation that humans cannot sense or even imagine.

Although the magnetic field is a useful tool when the stars are obscured by clouds, it isn't entirely foolproof either. Birds can become disoriented when traveling through magnetic anomalies such as areas with iron ore deposits. Scientist Charles Walcott found, for example, that some homing pigeons released within magnetic anomalies become confused until they manage to wander far enough away that they can orient normally. A magnetic compass alone would also be inadequate to guide birds across the equator. Because birds cannot distinguish between the North and South poles, a bobolink would become confused when reaching the equator unless it checked the magnetic field against another cue.

Instead of using a compass that indicates which pole is which, birds sense the angle at which magnetic field lines intersect the earth. The magnetic field is produced by the earth's molten iron core, as if a giant bar magnet is aligned toward the North and South poles. The magnetic field comes out at 90 degrees near each pole, wraps around the earth circularly, and is horizontal at the equator. By sensing where the field dips down toward the horizon, a bobolink knows it is facing a pole. If the field lines go above the horizon, the bobolink knows it is facing the equator. When a bobolink crosses the magnetic equator, the lines are parallel to the horizon. Then a bobolink must rely on another cue, such as stars when the sky is clear, to continue heading in the right direction. The next time it sees the lines dipping below the horizon—indicating that it is heading toward a pole—it somehow knows to switch from heading away from a pole to heading toward it.

When the bobolink can't rely on the stars or the magnetic field, it probably uses information from the setting sun to maintain its sense of direction. Experiments with thrushes by William Cochran and colleagues found that at sunset, just before the birds take off for a night of migration, they check the magnetic compass against the direction of the sunset and/or patterns of polarized light. Created when sunlight scatters as it passes through the atmosphere, polarized light patterns are invisible to humans. By using Polaroid filters to change the way caged birds perceive patterns of polarized light, researcher Ken Able showed that songbirds can use polarized light patterns as a directional compass.

In addition to the sun, stars, magnetic fields, and polarized light, birds probably use other cues to navigate. They can check their position against landmarks such as rivers and mountain ranges. Pigeons are also capable of smelling their way home. They apparently learn odors associated with particular wind directions. When they are released, they smell the air and fly away from the direction of wind carrying the odor. If far from home, they can also retrace their route by following the sequence of odors that takes them back from where they came.

At dusk, as a bobolink heads south from Maine hundreds of feet above the earth, it uses extraordinary senses to find its way. It sees the glow of sunset and patterns of polarized light, and it senses the magnetic field to pick an initial heading. During its flight, it can look up and see the reassuring pinpoints of light moving around the North Star.

It checks its progress against the shadowy mountains below, until it reaches the vast darkness of the Atlantic Ocean. Within its body, minute particles of magnetite are aligned with the poles. The bobolink feels the pull of the earth's magnetic field and flies onward in the darkness, miles beyond the prying eyes of scientists.

THE LONGER that Olin Sewall Pettingill studied bobolinks, the more he wondered about their lives on the south side of the equator. Although he consulted the ornithological literature and quizzed friends from South America, no one knew the answers to his most basic questions: Where exactly do bobolinks go in winter? What do they do there? When and where do the drab brown birds that departed Maine change into the striking black-and-white birds that arrive in spring?

In 1978, Pettingill and his wife set out for northeastern Argentina, where they guessed that rice fields on the lowlands flanking the Río Paraná might attract bobolinks. From Buenos Aires, they headed north, accompanied by local naturalist Maurice Rumboll. On the way, they passed through the dry cattle country of Santa Fe Province, where they found only a single bobolink. It was in a cage in the back of a shop lined with cages of saffron finches, red-crested cardinals, and other wild birds of Argentina. The shopkeeper said the bobolink had been caught and kept in order to lure other bobolinks to traps. The captured birds would be sold as caged birds for about four dollars per male or five dollars per pair. Remembering the bobolinks flying free above the grassy hills of Maine, Pettingill was distressed to see the bird imprisoned.

Because there was only one bobolink in the shop, Pettingill concluded that most of the bobolinks must be farther to the north. For the next three days, Rumboll and the Pettingills continued on past fields of sunflowers, sorghum, cotton, and rice crops. They saw many birds, but no bobolinks. One rice grower, James Cook, informed them that his crops were still immature so the "terrible pests" had not yet arrived to devour them. He offered to show them where another rice grower had planted an early crop. On January 13, Cook brought them to the edge of a vast rice field, where they got their first glimpse of bobolinks at home on their wintering grounds. Hundreds of bobolinks perched on the trees and bushes and on the fence. The cacophony of voices was so loud and discordant that Pettingill wondered whether they could be the same birds of North American meadows celebrated by poets.

The rice growers called the birds "charlatans" or "noisy ones." Cook said they damaged 8 percent of the crop. In some areas, farmers shot into the flocks to disperse them or set off cannons every fifteen minutes. Pettingill later returned to the seventy-five-acre rice field and watched as one flock after another rose up, circled, and disappeared again — thousands in all, making such a din that he could have heard them half a mile away. Over the next several days, the Pettingills explored the areas to the east and north, looking for bobolinks and finding them wherever rice was at the milk stage.

Five thousand miles from Maine, the bobolinks were at home among yellow-winged, chestnut-crowned, and scarlet-headed blackbirds. They shared the landscape with whistling herons and with southern screamers — odd birds that struck

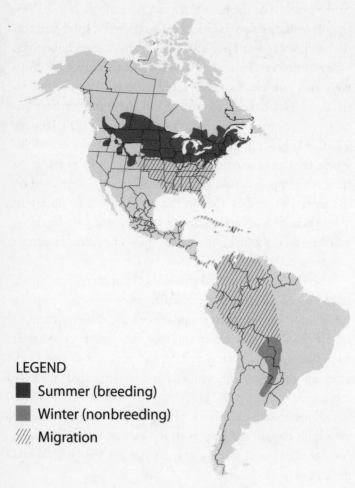

LEGEND
- ■ Summer (breeding)
- ■ Winter (nonbreeding)
- ///. Migration

Bobolink migration map.

Pettingill as having the bill of a turkey, the flight of a vulture, the posture of a cormorant, and the calls of a goose. He reflected on the peculiarity of New World birdlife. The bobolink, a North American bird, was passing its winter in the South American summer. In North America there was nothing comparable, since no South American birds ventured a visit north at any time of the year.

Back in Maine, the chill winter wind would be blowing across empty fields of snow. In the Argentinian summer, though, the resident birds were courting, mating, building nests, and laying eggs. The lengthening January days were already triggering the hormonal changes that were causing bobolinks to molt once again—Pettingill saw some males that were losing their feathers and growing new black and white plumage for spring. He also heard some of the males singing, as if warming up for courtship. But the longer days also brought on an inner restlessness that would keep the bobolinks from laying claim to a nesting place just yet. The right spot for a nest was waiting somewhere beyond the South American grasslands, the Caribbean Sea, and the rice fields of the southern United States. Their restlessness building, the bobolinks would take flight by darkness on January nights and, as suddenly as they had come, leave behind the varied birdlife of South America, swept irresistibly northward on an ancient tide.

9
LOOKING AND LISTENING FOR AUTUMN MIGRATION

And all about, the hills are crowned
  With woods that seem to burn and glow,
And purple asters, from the ground,
  Look up and watch the armies go;
Long, swaying ranks of swallows strong,
  And bobolinks, alert and gay,
And warblers, full of life and song—
  All moving swiftly on their way.

And silently, among the trees,
  The thrushes flock and disappear;
We hear their notes upon the breeze,
  And then—the singers are not here.
The autumn wanes, and kinglets go,
  Sweet-voiced and knightly in their way,
And all the birds our summers know,
  They flock and leave us day by day.
  —*Frank H. Sweet, from "Flocking of the Birds"*

S ONGBIRDS FLY SOUTH in autumn on a grand scale, their numbers augmented by millions of young birds that didn't even exist three months earlier. Although fall migration surpasses spring migration in sheer magnitude, it often receives less fanfare, perhaps because immature birds, and some adults, travel south in relatively drab plumage, without the embellishments of song. Despite these challenges — or perhaps because of them — many birders particularly relish autumn as a time to hone their bird-identification skills.

By paying attention to subtle variations in plumages and behaviors, birders tease apart the identities of "confusing fall warblers," birds whose autumn plumages present tricky problems for identification. They listen for the birds' call notes — simple vocalizations that can characterize some species as surely as their melodic songs. At night, they listen for migrants, tuning in to the ephemeral flight calls that can reveal whether Swainson's thrushes, black-and-white warblers, or dickcissels are passing overhead among the throngs of birds heading south in the dark.

Late summer is a time for transformation, as young birds fledge from the nest and the breeding season comes to an end. Male songbirds cease singing from the treetops and become more furtive, their melodies and conspicuous perching no longer needed for courtship and territorial defense. As the young become independent and roam about in their immature plumage, birders encounter more plain-looking birds than at any other time of year, making identification more difficult. Meanwhile, adult songbirds begin shedding their feathers, sometimes undergoing striking changes in appearance.

An adult male scarlet tanager, for example, is brilliant red with black wings in summer, but yellowish-green with black wings in autumn. On its wintering grounds, it changes plumage again and comes north in red and black. A mature male blackpoll warbler has a distinctive black cap and bright white cheeks in spring, but the black disappears with the autumn molt, replaced by olive feathers that blend with the new yellowish wash on its head and face.

Adult males of about ten northern American warbler species have different plumage patterns in spring than in fall, but most of them retain familiar field marks year-round that provide clues for identification. For example, a male chestnut-sided warbler has distinctive chestnut streaks on its sides all year, despite the change in color on its face and back. Similarly, a male bay-breasted warbler's flanks always show some rufous, even though its face and head look completely different — black, yellow, and reddish-brown in spring, greenish-yellow in autumn.

Although autumn birding is notorious for "confusing fall warblers," most songbirds actually migrate in colorful glory both in autumn and in spring. About forty warbler species appear the same all year long, or are slightly muted in color, but still recognizable. A southbound male American redstart is dazzling, feathered in orange and black. An ovenbird has dark streaks on its breast and an orange crown in autumn as in spring. A male black-throated blue warbler is showy throughout the seasons, with a black face, a snowy white belly, and a dark blue back.

With young birds, though, the real confusion begins — not only because there are new plumages to learn, but also because immature birds have fewer notable field marks and can look maddeningly similar to the females or the year-lings of other species. For example, a black-and-orange Blackburnian warbler and a blue-and-white cerulean warbler are unmistakable, but the immature birds are nondescript. Both are yellowish below and grayish above and have a broad yellowish line, or supercilium, above the eye. Both have two white wingbars and a plain breast with some streaking along the sides. After narrowing down the possibilities to those two species, however, birders look for the telltale aqua-green tint to the cerulean warbler, or the pale lines on a Blackburnian's back and the pointed pattern of the gray feathers that cover its ear. Distinguishing a young bay-breasted warbler from a blackpoll warbler requires knowing a different suite of clues such as subtle differences in color pattern: blurry streaks on the flanks and backs of the blackpoll warbler and yellow, rather than yellowish-gray, soles of the feet (if one is lucky enough to catch a glimpse).

Learning to recognize new and subtle variations in fall plumages can be daunting, but several steps can aid the effort. Use a field guide that illustrates the plumages of birds of different ages at different times of the year. It may help to focus on common migrants and to note when they generally pass through. Yellow warblers are one of the earliest migrants, gone from most of the northern United States by late September, but yellow-rumped warblers move south well into October and November.

In addition to studying plumage patterns, notice how the birds behave. Magnolia warblers flutter actively in the trees, spreading their tails to flush prey out of hiding. Immature prairie warblers may resemble immature magnolia warblers, but prairie warblers bob their tails and forage on the ground or at the forest edge. Listening to the birds can also help clinch identifications. Even though migrants don't usually sing in autumn, they utter simple call notes that are often distinctive. For example, yellow warblers and common yellowthroats may appear similarly drab olive with yellowish undertail coverts, but a yellowthroat utters a dry *tchat,* and a yellow warbler calls out with a clear-sounding *chip.*

IN A VENTURE that has become known as the cutting edge of birding, some experts have learned to identify songbirds as they fly past based on general color patterns, size, structure, flight style, flight calls, and a general knowledge of when each species is likely to be seen. Naturalists from the Cape May Bird Observatory conduct a project called Morning Flight that identifies and counts songbirds during brief daytime flights during autumn migration. It's similar to a hawk watch except that observers must be able to identify about ninety species rather than a dozen. Moreover, the birds are only four to five inches long, rather than two to three feet, and they zip past in a matter of seconds, rather than soaring slowly above.

Beginning at sunrise from September 1 to October 31, the Morning Flight count takes place from atop a massive dike at the Higbee Beach Wildlife Management Area on the tip of Cape May peninsula in New Jersey. By luck of geography, it's one of the best places in the world to see songbirds migrating by daylight at close range. In fall, when the birds ride a cold front southward all night long, some drift to the coast or offshore and decide to come back toward land. Beginning around sunrise, songbirds stream north past

Higbee Beach, apparently seeking better migration routes or feeding areas for the day.

On a typical September morning, Michael O'Brien, one of the regular counters, is standing on top of the dike with a full view of the sky. When a speck of a songbird flits past just above eye level, he marks a tally for yellow warbler — stocky, broad-winged, and bright yellow from the chin to the tip of the tail. When another tiny bird streaks by a minute later, he records northern parula — a relatively small warbler, round bodied and short tailed, with a bouncy flight. Then he spots a lone bird flying above the dike and records it as a "baypoll," a bay-breasted warbler or a blackpoll warbler. All the while, he listens for high-pitched flight calls, the same calls that the birds use when they migrate at night. A high-pitched *tszp* from a bird passing high overhead gives it away as a Cape May warbler.

On a slow morning with a southwest breeze, O'Brien might count only a few dozen birds. If the winds shift and come from the north or northwest, he could see as many as several hundred to several thousand birds the very next day. Increasingly favorable winds could bring in more than nine thousand birds the morning after that. Numbers vary a lot depending on the season and the weather, and the observatory often sends two people up on the dike to keep track of big flights. The slowest mornings may lack migrants altogether, but high counts have exceeded one hundred thousand songbirds. These massive movements tend to occur late in fall, when short-distance migrants, such as yellow-rumped warblers and American robins, fly past by the tens of thousands.

By watching morning flights, O'Brien says he has

learned a lot about the influence of weather and the changing seasons on songbird migration, patterns that are more difficult to perceive when watching songbirds after they have landed. Eventually, researchers will compare the counts with recordings from monitoring stations in New Jersey, Pennsylvania, and Delaware that automatically record the flight calls of migrants throughout the night. The data could also be compared with radar images to better understand the magnitude and composition of migratory flights on any given night.

LISTENING TO NOCTURNAL flight calls is a powerful way to appreciate migrations passing overhead in the dark. Unlike radar, it offers clues about the identities of the birds, and unlike morning flights, which occur primarily along the coast, anyone can tune in to migration by listening at night from just about anywhere. This pastime has become increasingly popular among serious birders ever since Bill Evans and Michael O'Brien published a comprehensive CD-ROM of nocturnal flight calls in 2002. Nocturnal flight calls are also showing promise as a way to monitor bird populations for conservation, now that computers can assist in detecting and sorting the calls of different migrant species.

As recently as 1985, that capability was a distant dream. People knew that migrants called from overhead at night; when Swainson's thrushes passed over, it sounded as though the calls of spring peepers were coming down from the sky. In addition to the thrushes, other birds were uttering a variety of high-pitched, fleeting calls, but their identities were unknown. That began to change after Bill Evans, a

University of Minnesota student, had an epiphany while listening to nocturnal migrants.

After delivering pizzas one evening, Evans had camped out on a bluff along the St. Croix River to maximize his chances of watching migrants at dawn. As he lay awake, he heard what he thought was the call of a black-billed cuckoo. Listening intently, he counted more than one hundred cuckoos passing overhead in a single hour — an astounding number, given that cuckoos are secretive by daylight and are hardly ever seen. Suddenly he realized that recording the calls could be a powerful new way to monitor birds for conservation.

During the late 1950s, Richard Graber and William Cochran had recorded the calls of migratory birds at night, hoping to learn more about the numbers and kinds of birds passing over. They were impeded by two major obstacles: the difficulty of distinguishing different species' calls, and the laborious and time-consuming task of listening to tapes. For Evans, the effort would turn into a life's pursuit.

Beginning in 1985, Evans recorded and listened to thousands of hours of tapes. He documented hundreds of call variations and began matching each one to the species that uttered them. In 1990, he produced a cassette tape that identified the flight calls of the *Catharus* thrushes, including wood thrush, veery, Swainson's thrush, gray-cheeked thrush, and hermit thrush. One of the tapes made it into the hands of Michael O'Brien, who had been fascinated by night flight calls since his boyhood. In his Rockville, Maryland, neighborhood, he had listened to thrushes and other migrants passing overhead before dawn as he delivered newspapers. At night, he sat in his front yard, listening to the birds overhead, but he hadn't realized that anyone else had the same interest. After hearing the thrush tape, he called Evans, and the two realized instantly that they shared a rare passion and enthusiasm for listening to nocturnal migrants.

In 1991, Evans and O'Brien began collaborating to identify the remaining unknown calls of eastern migrants, and their passion would be a key to their success. Although Evans had already sorted out the calls of about seventeen warblers, the remaining ones were progressively harder to identify, often requiring painstaking detective work. Since most songbirds utter their flight calls by day, especially soon after landing in the morning, Evans and O'Brien recorded the vocalizations of migrants they identified visually by daylight to match them up with unidentified calls recorded at night. Night sessions from Florida, for example, had recorded a buzzy note with a rising pitch more than a thousand times, but its caller remained unknown until 1993, when O'Brien captured on tape the sound of a northern waterthush in Virginia. When the two squiggles on the sonograms matched, the mystery was solved instantly.

Other calls fell into place fortuitously. By 1995, only a few unconfirmed calls remained, two of which had been narrowed down to either Wilson's warbler or Canada warbler. When a friend sent Evans a recording from Vancouver Island that included one of the mystery vocalizations, a liquid-sounding *spiv* call, he realized that it must be Wilson's warbler, since the Canada warbler doesn't occur that far west. Later he unexpectedly pinned down the other call when friend Lang Elliott recorded the song of a Canada warbler and played it for Evans. Mixed in with the song was a familiar liquid *pcht* note. Evans excitedly ran the call through his computer and found that the note perfectly matched the V-shaped spectrogram of the last unknown call. After more than a decade of effort, Evans and O'Brien produced the landmark guide that identified the flight calls of 211 species of migratory landbirds in eastern North America.

The guide was an essential step to making the acoustic monitoring dream come true, but so was developing the technology to automate sound recording and analysis. In 1994, Evans began collaborating with researchers and software programmers at the Cornell Laboratory of Ornithology who had developed computer software to pick out the sounds of whales from hundreds of hours of undersea recordings. By 2000, Cornell Laboratory of Ornithology researchers and collaborators were monitoring the night sky from the rooftop microphones of nine homes in the Delaware Valley. Cables connected the microphones to computers inside that kicked on automatically each night and continuously recorded the night sky until dawn. A software program flagged sounds with acoustic properties similar to the ones biologists were looking for. The data were sent automatically by modem to the Cornell Lab in Ithaca, New York. There, a technician examined the data, rejecting background noises such as raindrops and katydids and extracting the calls of warblers and sparrows. In this way, the number of calls from several stations could be categorized and summarized within a few hours rather than the fifty hours it would have taken for a human to listen to the tapes unaided.

At least thirty stations across the country are using this technology to record nocturnal migrations. Eventually these stations could be networked and standardized, providing new insights and adding depth to existing bird-monitoring efforts. Acoustic monitoring is especially valuable for detecting species such as Canada warblers, Wilson's warblers, and gray-cheeked thrushes that inhabit forests beyond the reach of the Breeding Bird Survey, which extends only as far north as there are good roads to travel.

As Evans envisioned the first night he heard one hundred black-billed cuckoos, acoustic monitoring is also a way to survey birds that are so cryptic that they're difficult to census in any other way. Additionally, recordings of vocalizations are providing new information about regional and even local differences in migration routes. For example, by recording calls from across New York state, Evans learned that gray-cheeked thrushes migrate in greater numbers through the western part of the state, whereas rose-breasted grosbeaks migrate in greater numbers to the east. Monitoring stations are also helping to document the passage of migrants at specific sites to aid in decisions about placing communications towers and wind farms in areas that minimize harm to migrating birds. In the long term, acoustic records from across the continent could provide an archive of data to detect changes in the numbers of migrants through the years.

Many of these stations could be run by bird watchers from their own homes. Using instructions posted on Evans's Web site, anyone can rig up a rooftop microphone for about thirty dollars, using a plastic dinner plate to help amplify the sound and a flowerpot to house the mike. A cable connects the microphone to a computer inside the house. With this setup, a person can record the calls overnight and wake up in the morning to explore what passed overhead.

This setup has also provided a new way for birders to witness the awesome migrations they wouldn't otherwise be able to see. In Ithaca, New York, bird migration buffs sometimes gather at the home of birders Chris and Diane Tessaglia-Hymes on cold autumn nights. In the hushed room, they sit around the television, listening for the high-pitched calls of birds, relayed from the rooftop microphone through the television speaker, and watching as real-time sonograms scroll across the screen. When they hear a raspy *vheer* and see a long flat arch on the sonogram, someone calls out, "Gray-cheeked thrush!" If they hear a delicate *dzinn* and see a drawn-out sonogram resembling the shape of an inchworm, they know a black-and-white warbler has just passed over the house. A whistled *puwi* with a thin line of a spectrogram means that a rose-breasted grosbeak has flown by, and a long sweet *tseedt* means a white-throated sparrow is winging its way beneath the stars. Sometimes Chris Tessaglia-Hymes stays up well into the night, listening to the birds passing overhead. Part of the fascination, he says, is in listening to hundreds of migrants and never knowing what bird might be coming around the bend. The sheer numbers of birds can also be thrilling. In a single night, he can sometimes hear more gray-cheeked thrushes passing by than most people would ever see in a lifetime.

No matter where you live, migratory songbirds are likely to be passing through the air just hundreds or thousands of feet above where you sleep. Go outside on an autumn night, and you might hear the stream of migrants, invisible overhead. Evans says that the sounds of nocturnal songbirds stir something in him, the same way that migrating geese do when he hears them traveling south in the fall, or when he hears a flock of oldsquaw coming off Lake Ontario on their journey from the Arctic to the Chesapeake Bay. On a night in August, he might hear Canada warblers call from above, their tiny bodies hurtling over upstate New York on their way from Canadian forests to the Andes in Peru. The delicate, upslurred *tsweet* calls of American redstarts drop down from the sky, a last farewell as they head for the Caribbean. Bobolinks from the grasslands of Canada utter their metallic *clink* notes on their way to the pampas in Argentina. Evans says these sounds open up the vastness of migration in his mind. "It's as if anyone can tune into the continents," he says, "just by listening right from where they are."[1]

Hot Spots for Watching, Listening, and Learning About Autumn Migration

Watching songbirds in autumn can be especially revealing in places that attract large numbers of migratory birds. Many bird observatories, bird-banding stations, and nature centers offer banding demonstrations and bird walks that allow birders to get close-up looks at warblers and other birds in the hand or to learn from experts in the field. Places with outstanding interpretive programs and fall songbird migrations include Cape May Bird Observatory in New Jersey, Long Point Bird Observatory in Canada, PRBO's Palomarin Field Station and the Big Sur Ornithology Lab in California, the San Pedro National Riparian Conservation Area in Arizona, and Fort Morgan in Alabama with the Bird Banding-Hummer Bird Study Group.

*Cape May Bird Observatory's Morning Flight Counts
at Higbee Beach, Cape May, New Jersey*

As birds move south through New Jersey along the Atlantic Coast in autumn, they funnel down the Cape May
peninsula, renowned for its concentrations of migratory
birds in spring and fall. When planning a trip to watch migratory songbirds in Cape May, be prepared for action.
Wear loose clothing and leave your 10x binoculars at home,
and be ready to swivel at the hip, advises Pete Dunne, vice
president of Natural History Information at the Cape May
Bird Observatory. You won't want anything to restrict your
movement, your field of view, or your ability to focus on
tiny birds that are moving quickly.

The Higbee Beach Wildlife Management Area is an especially good place to look for songbirds in transit. During
peak migration when a cold front blows through from the
north, millions of migrants ride the winds south. At daylight, when some of those birds find themselves offshore,
thousands may stream back northward, passing over Higbee Beach as they seek better places to land. Warblers con-
stitute the bulk of the migrants, along with vireos, tanagers,
grosbeaks, and other songbirds, the composition changing
throughout the season.

Researchers from the Cape May Bird Observatory conduct the Zeiss Sports Optics Morning Flight songbird count
from atop a dike near the parking lot. On some mornings,
interpretive naturalists sit atop a nearby observation platform and are available to help visitors identify warblers zip-
ping past and to answer questions about the research
program.

On the ground, one of the best places to see songbirds is

in the shelter of trees near the parking lots and fields, especially on days when cold fronts are blowing through. Look for songbirds along the trails, especially at the edge of the dune forest. On good days, Higbee Beach teems with birds. On different record-breaking days, observers at Higbee Beach estimated counts of astounding numbers of birds, including 100,000 yellow-rumped warblers, 1,000 rose-breasted grosbeaks, 1,000 black-throated blue warblers, and 550 red-eyed vireos.

Surrounding areas are also good for witnessing songbird migration. The Rea Farm is a great location, particularly on cold mornings with northwest winds, when birds take shelter in the woodland pockets. Check the hedgerows and forest of Hidden Valley Ranch and the brushy areas around the Cape May Bird Observatory. Cape May Point also yields numerous songbirds on migration. On a record-breaking day for bobolinks at Cape May Point, observers estimated there were fifteen thousand bobolinks.

WHEN TO GO: Beginning of August to second week in November. The best times to watch warblers are the last week in August to the first week of October. The Morning Flight project takes place from September 1 through October 31 for four hours each morning, beginning at sunrise.

NEAREST BIG CITY: Cape May is about fifty miles southwest of Atlantic City.

SPECIAL ACTIVITIES: The Cape May Bird Observatory offers birding workshops in the field and classroom, including topics such as fall migration, warbler identification challenges, and identifying birds on the wing. Pete Dunne is a frequent leader of walks and workshops, and Michael O'Brien has taught a workshop about birding by ear. Registration fees are required. For more information, call (609) 861-0700 or visit http://www.njaudubon.org/Calendar.

FOR MORE INFORMATION: Cape May Bird Observatory Center for Research and Education: (609) 861-0700; http://www.njaudubon.org/Centers/CMBO/#Goshen.

New Jersey Audubon Society: http://www.njaudubon .org.

RECOMMENDED READING: *The Birds of Cape May,* by David Sibley. Bernardsville: New Jersey Audubon Society, 1997.

Wild Journeys: Migration in New Jersey, by Brian Vernachio, Don Freiday, and Dale Rosselet. Bernardsville: New Jersey Audubon Society, 2003.

Long Point Bird Observatory's Old Cut Banding Station, Long Point, Ontario

Jutting out nearly twenty miles into Lake Erie, Long Point is known for its concentrations of songbirds during spring and fall migrations. The Long Point Bird Observatory, established in 1960, is North America's oldest bird observatory. Over the years, more than 800,000 birds of about 275 species have been banded at Long Point. Visitors to the Old Cut banding station are welcome to watch as staff and volunteers capture, band, and release birds for scientific studies. On a ten-minute stroll along the narrow paths where the mist nets are hung in a grove of trees, visitors may see American redstarts and bay-breasted, chestnut-sided, magnolia, and black-throated blue warblers in the nets. Half an hour later, the nets might hold ovenbirds and black-poll, yellow-rumped, and Canada warblers. The one-room banding station offers a few chairs for visitors and enough standing room for about half a dozen people (though more may cram in during the peak part of the season). The banders will explain what they are doing as they handle and measure the birds and share stories about what they have learned from long-term monitoring.

Some 378 species of birds, including 38 warbler species, have been found at Long Point. The areas in and around Long Point Provincial Park are good for warbler watching, especially the small pine plantation on the road to the right just inside the park. Migratory songbirds also pass through in large numbers along the causeway willows by the Big Creek National Wildlife Area. Bird Studies Canada's headquarters in nearby Port Rowan is surrounded by thirty-two acres that are accessible by trails. During migration, songbirds sometimes move through the trees along the shoreline.

WHEN TO GO: Spring migration peaks in May; autumn migration peaks in September. The banding station and visitor center are open to the public from April to mid-June and early August to mid-November. Typical hours are eight A.M. to noon. Because the banding schedule may vary depending on weather or other factors, the observatory advises calling (519) 586-2885 in advance.

NEAREST BIG CITY: Long Point is about one hundred miles southwest of Toronto.

SPECIAL ACTIVITIES: Long Point Bird Observatory offers tours of the on-site monitoring program for schools, universities, naturalist clubs, and birding groups of up to thirty people. Advance reservations are recommended.

RECOMMENDED READING: *A Birding Guide to the Long Point Area,* by J. Skevington, B. Collier, and T. Woodrow. Port Rowan: Bird Studies Canada, 1990. The book is available at the Bird Studies Canada headquarters or by calling (888) 448-BIRD.

FOR MORE INFORMATION: Long Point Bird Observatory: (519) 586-2885; http://www.bsc-eoc.org/lpbo/lpbirdo.html.

PRBO's Palomarin Field Station, Point Reyes National Seashore, California

Located on the Pacific Coast just north of San Francisco, the Point Reyes National Seashore is an excellent bird-watching destination year-round. More than 45 percent of North American bird species have been found in diverse habitats within some one hundred square miles, including estuaries, beaches, coastal scrub grasslands, riparian areas, salt and freshwater marshes, and Bishop pine and Douglas fir forests. Although Point Reyes doesn't receive the high concentrations of migratory songbirds that eastern hot spots do, local spots within the park are good for spring and fall migrants.

PRBO Conservation Science (formerly the Point Reyes Bird Observatory) runs the Palomarin Field Station, which has a visitor center that is open to the public from sunrise until five P.M. The field biologists usually keep the mist nets open for six hours each day beginning fifteen minutes after sunrise. The staff also holds bird-banding demonstrations on certain days of the week. Since nets may be closed on cold or windy days, call (415) 868-0655, extension 315 to check the schedule and conditions. In fall, common migrants include warbling vireos; Swainson's thrushes; black-headed grosbeaks; ruby-crowned kinglets; and orange-crowned, MacGillivray's, and Wilson's warblers. Look for birds on nearby trails, including the Arroyo Hondo trail that winds along a creek through various stages of lush vegetation.

The twelve-mile-long Point Reyes peninsula extends into the Pacific Ocean. The tip, near the Point Reyes lighthouse, is a good place to look for migratory songbirds, especially vagrants. In between the parking lots and the lighthouse, migrants land in a row of gnarled Monterey pines. On cool, overcast days in autumn, birders may find black-headed grosbeaks, varied thrushes, and as many as eighteen species of wood-warblers, including Townsend's and MacGillivray's. Rarities such as prothonotary warblers, yellow-throated vireos, gray-cheeked thrushes, and lark buntings sometimes show up here, hundreds or thousands of miles off course. In the open grassland and coastal scrub, migratory birds concentrate in Monterey cypress trees planted near ranches as windbreaks. The willows near the parking lot at Drake's Beach and the cypress trees near the parking lot at the Chimney Rock trailhead are good places to check.

Riparian areas attract the greatest diversity of songbirds, along the edges of creeks in willow and alder trees. In particular, check the lush riparian jungle at Olema Marsh, the riparian area along the Earthquake Trail near the Bear Valley Visitor Center, and the willows along Five Brooks Pond.

WHEN TO GO: In spring, mid-March to the end of April for most migrants, May 25 to June 15 for vagrants — birds wandering beyond their usual range. Mid-August through October for fall migration. Peak songbird diversity occurs around the third week of September.

NEAREST BIG CITY: Point Reyes is about thirty miles northwest of San Francisco, approximately a one-hour drive.

SPECIAL ACTIVITIES: The Point Reyes National Seashore Association offers periodic field seminars for a fee, including seasonal topics such as "The Pageant of Migration," led by Rich Stallcup. For information, call (415) 663-1200 or visit http://www.ptreyes.org/field/fshome.html.

FOR MORE INFORMATION: PRBO Conservation Science: (707) 781-2555; http://www.prbo.org.

Point Reyes National Seashore: (415) 464-5100; http://www.nps.gov/pore.

RECOMMENDED READING: *The Natural History of the Point Reyes Peninsula,* by Jules G. Evens. Point Reyes Station, Calif.: Point Reyes National Seashore Association, 1993.
Field Checklist of Birds for Point Reyes National Seashore, by Richard W. Stallcup. Point Reyes Station, Calif.: Point Reyes National Seashore Association, 2000.

WINTER

10
*B*IRDS OF TWO WORLDS

In Cuba most of our wood warblers are known simply as
"Mariposas" — butterflies; but the redstart's bright plumage
has won him the name "Candelita" — the little torch that
flashes in the gloomy depths of tropical forests.
— *Frank M. Chapman, from* Handbook of Birds

IN THE MOUNTAINS OF the Dominican Republic, the voices of tropical birds rise from the forest in the darkness before dawn. The punctuated songs of La Selle's thrushes resound from the trees, *TWEE-ker, TWEE-ker, TWEE-ker,* intermingled with the haunting whistled notes of rufous-throated solitaires. Slowly the forest becomes visible in the gray light — thick tangles of vines and enormous, shadowy tree trunks rising up to a gnarled canopy resembling that of a towering bonsai garden. Black-headed palm tanagers flutter among the vines, and a sparrow-sized Hispaniolan spindalis hops in the brush. Unseen, another bird calls insistently, *pweer, pweer, pweer,* until the sun begins trickling down through the trees and the voice vanishes. In darkness each morning, the sound begins again, then stops soon after dawn. At last, when the dry season comes, the *pweer* drops out altogether from the tropical chorus among the forest of *palo de vientes,* trees of the wind.

Two months later and thousands of miles to the north, the voice resumes in nearly unrecognizable beauty on another mountaintop. Likened to a song breathed into a delicate golden tube, the melody begins with a soft *chook-chook,* followed by a series of ethereal notes that spiral downward, then rise at the end. In 1881, twenty-one-year-old Eugene Bicknell heard the unfamiliar song on an expedition to the mist-shrouded Catskill Mountains in New York. When the bird flew into view, Bicknell quickly raised his gun and shot it down. The bird, limp and warm in his hands, had an olive-brown back and a buffy breast with dark spots. Later, ornithologist Robert Ridgway examined the specimen and named it Bicknell's thrush, a subspecies of gray-cheeked thrush.

For more than a century, Bicknell's thrush seemed destined to remain an obscure subspecies. Then, in 1991, researcher Henri Ouellet spoke at a meeting of the American Ornithologists' Union, presenting evidence that the outward similarities between Bicknell's and gray-cheeked thrushes had been deceiving. When Ouellet played recordings of their songs, Bicknell's and gray-cheeked thrushes ignored one another's vocalizations. Moreover, differences in DNA sequences showed that Bicknell's and gray-cheeked thrushes had been genetically separated for one million years. A careful examination of specimens revealed that Bicknell's thrush was smaller, was browner on its back, and had a tinge of chestnut in its tail.

The two thrushes also inhabited distinct regions, both in summer and in winter. The gray-cheeked thrush bred in a broad swath of taiga from Alaska to Newfoundland and wintered in the Amazon rainforest. Bicknell's thrush spent its summers among the isolated mountaintops of New York and New England in the United States, and Nova Scotia, Quebec, and New Brunswick in Canada. In winter, Bicknell's thrush inhabited just a few islands in the Greater Antilles.

When biologist Chris Rimmer heard Ouellet's talk, he began thinking about Bicknell's thrush, one of the most enigmatic birds in all of North America. Whether or not it was a separate species, it seemed that conservationists ought to find out more about it, and the opportunity was essentially in his backyard. Rimmer, director of conservation biology at the Vermont Institute of Natural Science, began looking for Bicknell's thrushes in Vermont and nearby states with the help of volunteers. In 1992, he led a research team to Vermont's Mount Mansfield to learn more about their poorly documented nesting habits. Three years later, the American Ornithologists' Union declared Bicknell's thrush a new species, bringing international attention to its uniqueness and the need for its conservation.

Rimmer's team quickly learned why Bicknell's thrush was one of the least known birds in North America. On the Vermont mountaintops, the researchers were taller than the stunted firs, but the miniature forest was so thick that they could hardly see five feet in front of themselves. To survey the rugged area, they had to push the soft branches aside as if swimming through a sea of trees. By day, the fog and dense, dark green vegetation concealed the birds' movements. At dawn and dusk, the thrushes uttered their flute-like songs, tantalizingly close, but still invisible. After capturing a Bicknell's thrush in a mist net and releasing it with a band, they rarely saw it again, unless they happened to capture it once more.

No one knew how many Bicknell's thrushes existed. Estimates ranged from twenty-five thousand to fifty thousand. It was especially worrisome that the birds were confined to scattered mountaintops totaling roughly 740 square miles — less than half the area of Rhode Island. Their wintering area was even smaller — primarily in the Dominican Republic, with just 43 square miles of good habitat and about 122 square miles of marginal habitat left. Rimmer's team knew they had to act quickly to find out exactly where the birds wintered and to advocate for their protection.

On December 1, 1995, Rimmer and two field biologists, Kent McFarland and Jim Chace, embarked on an expedition to the Sierra de Bahoruco in the southwestern Dominican Republic. After a three-hour drive up the mountain over rutted dirt roads, they arrived at one of the island's last remaining ridgetop forests. With Hispaniolan crossbills, western chat-tanagers, and La Selle's thrushes in the imposing forest of *palo de vientes*, they seemed a world apart from the firs of Mount Mansfield. Just about the only similarity was the dense vegetation, except that they needed a machete to cut through the tropical vines. As they cleared the vines to make a narrow lane for mist nets, they also heard a familiar sound: the slurred call notes of Bicknell's thrushes.

Early the next morning, McFarland and Chace awoke in the dark and hiked up the steep slope. They would have only about a fifteen-minute window before sunrise to capture the thrushes. After that, the birds would become too wary to catch. As McFarland and Chace had often done on Mount Mansfield, they unfurled the mist net, then retreated to a spot nearby and crouched down. McFarland pressed the "start" button on a portable tape recorder. *Pweer, pweer.* At the sound of the recording, a nearby thrush began calling back. *Pweer, pweer, pweer.* Agitated by the presence of the phantom intruder, the bird suddenly darted into the net.

Elated, McFarland and Chace ran up to the net. In the light of their headlamps, they found, to their surprise, that the thrush already had a band on its leg. After untangling the bird, they read the number: 1231-40012. Then Chace exclaimed, "I remember that combination. We banded this bird on Mount Mansfield!"[1] Skeptical, McFarland suggested that Rimmer must have banded it the previous winter during a scouting trip to the Sierra de Bahoruco, but Chace kept insisting. As soon as they got back to base camp, they looked up their records. The thrush indeed had been banded on Mount Mansfield six months earlier, on June 16.

They were dumbfounded. The odds were unfathomable. The first bird that they captured in the Dominican Republic that winter just happened to be one that they had held and banded on the slopes of Mount Mansfield — 1,800 miles away. McFarland likened it to winning the lottery and being struck by lightning at the same time. For the next week, McFarland and Chace often heard the thrush at the same spot but they were unable to capture it again. The following summer, though, 1231-40012 turned up in mist nets on Mount Mansfield, and then again the summer after that.

The fact that they had held the same thrush on either end of its migratory journey brought home the direct connection between the mountains of Vermont and the Dominican Republic. About 92 percent of the Dominican Republic's forests are gone, cleared for farming, homes, and lumber. Across the border in Haiti, the eastern side of the island of Hispaniola, there is virtually no forest left. If the remaining forests of Hispaniola are destroyed, Bicknell's thrushes might never make it back to Vermont. Similarly, if the montane forests of Vermont and nearby regions should ever disappear, Hispaniola might lose the Bicknell's thrush, too.

The impenetrable fir-covered slopes of Vermont are not in immediate danger. For the most part they're protected, or are simply too rugged and remote to be much good for anything other than ski trails and wind farms. Fortunately, Bicknell's thrushes seem unperturbed by skiiers and wind turbines, as long as there is enough forest on either side to inhabit. However, there are other potential threats, foremost among them the death of forests from global warming.

The balsam fir and spruce forests are on mountaintops for a reason: they thrive in the cold air of high elevations. As increasing temperatures creep up the mountainsides, the trees will have nowhere to retreat. Preliminary projections show that if global temperatures increase by about five degrees, there could be virtually no habitat left for Bicknell's thrushes in the United States. Meanwhile, the health of northeastern forests has already been compromised by acid rain, which makes spruce trees more susceptible to freezing and depletes calcium from the soil that is needed by invertebrates and the birds that feed on them. Mercury in the rainfall has also contaminated the forest and the blood running through the veins of thrushes. Conservationists worry about the combined effects of these stresses on the birds and their young.

Already, Bicknell's thrush is one of the rarest migratory birds in North America. It probably always was, given its restricted range and penchant for forested islands in the sky and sea. It will probably also remain as one of the most obscure simply because it is hardly ever seen. Its predicament, however, is shared by many birds that are more widespread and numerous — birds that anyone in eastern North America can see in a wooded park on a good spring day. Black-throated blue warblers, Cape May warblers, prairie warblers, palm warblers, black-and-white warblers, and ovenbirds spend the winter in Hispaniola's ridgetop forests, too. Several of these species are largely restricted to the Greater Antilles in winter — a chain of islands where most of the forests have already been cut down and deforestation continues. The birds are our connection to the Caribbean, and we're snipping off their lifeline on both ends.

OF COURSE, the connections extend all across Central and South America, too. As migrants flood southward from the broad expanses of the United States and Canada, their numbers become especially concentrated in the southern United States, a smattering of Caribbean islands, and Mexico. Biologist John Terborgh estimates that more than half of all Neotropical migrants are packed into about 1.4 million square miles in Mexico, the Bahamas, Cuba, and

Hispaniola—about one seventh the area of the United States and Canada north to the treeline. This is an especially crucial region for bird conservation, since every acre of forest shelters more migratory birds than the equivalent area on the breeding grounds.

On their wintering grounds, migrants converge from many parts of North America. Western breeders—such as black-throated gray warblers, MacGillivray's warblers, and western tanagers—tend to winter in Mexico and/or Central America. Many birds that breed in eastern North America winter primarily in the Caribbean, but others overlap with western species in Mexico and Central America. Cliff swallows, purple martins, veeries, blackpoll warblers, and Canada warblers are among the few species that travel exclusively to South America.

Though ornithologists have mapped the distributions of migrant species from localities written on the tags of museum specimens and from expedition records, they know next to nothing about where any particular bird ends up for the winter. Will a black-and-white warbler from the coniferous forests of Canada spend its time on a coffee plantation in Cuba, a patch of rainforest in the Campechean lowlands of Mexico, or an evergreen cloud forest in the Peruvian Andes?

Rare recoveries of banded birds offer small but important glimpses of where migrant birds go in winter. One Swainson's thrush was banded in British Columbia in June 1991, then captured in Arizona later that September, and in El Salvador in December the year after that. A cliff swallow banded by researchers Charles and Mary Brown in Keith County, Nebraska, in July 1990 was found by Jose Alberto

Nogueira six months later in Santa Helena, in southwestern Brazil, some five thousand miles away.

Birds from the tropics have also carried their bands thousands of miles north. In February 1989, researchers Barbara Dowell and Chandler Robbins captured an ovenbird on a cacao plantation near La Democracia, Belize, and released it with band number 2061-96079. Eight years later on a June day, Becki Michalak found the same ovenbird dead outside her home in Saegertown, Pennsylvania. It was one of the oldest ovenbirds on record. A Kirtland's warbler banded on Eleuthera Island in the Bahamas was resighted on its territory in Michigan — an especially valuable recovery since Kirtland's warblers are endangered, with only 2,100 estimated to remain.

Recoveries of multiple birds from the same location show that for any particular species, individuals from far-flung regions mix on the wintering grounds. During the 1930s and 1940s, more than 375,000 chimney swifts were banded in an effort to learn where the birds were wintering. In 1944, the United States bird banding office received word that thirteen of the banded swifts had been killed by Indians along the River Yanayacu in Peru. Five of the swifts had been banded by a Boy Scout troop in Memphis, Tennessee; the others had come from Ontario, Connecticut, Illinois, Alabama, and Georgia.

In the 1990s, scientists began looking for clues about bird migration, based on the molecules in their blood and feathers. Basic elements, such as nitrogen, carbon, and hydrogen, are found in different forms, or isotopes, that differ by one or more neutrons. The mix of isotopes varies depending on geography and habitat. When birds consume

these isotopes in their food, the isotopic "signatures" become incorporated into their bodies. The signature in a
bird's feathers reflects the ratio of hydrogen isotopes where
the bird molted and grew its feathers—usually assumed to
be an area close to the breeding grounds. Carbon isotopes
in a bird's blood can provide clues about where it has been
within the last six months.

By analyzing the birds' feathers, researchers found that
several species in Guatemala—among them the worm-
eating warbler, hooded warbler, wood thrush, gray catbird,
and ovenbird—had come from throughout their summertime ranges. In other cases, stable isotopes have helped reveal geographic patterns in migration. For example, isotopic
signatures suggest that black-throated blue warblers from
the northern part of their breeding range tend to winter on
the westerly Caribbean islands, Cuba and Jamaica. Those
from the southern United States winter on the easterly islands of Hispaniola and Puerto Rico. These findings could
explain why black-throated blue warblers in the southern
United States are declining more quickly than those in the
northern portion of the range. It may not be a coincidence
that the most extensive deforestation in the Greater Antilles
has occurred on Hispaniola.

BEFORE THE 1980s, most ornithologists viewed the world
of Neotropical migrants through a long-distance lens. They
read descriptions of exploratory expeditions such as the
ones led by Alexander Wetmore to South America in
1920–21 and to Panama during the 1940s–60s. They pored
over accounts of birdlife in Central America by pioneering
ornithologists, including Alexander Skutch in Costa Rica,

James Bond in Jamaica, and Paul Schwartz in Venezuela. They pulled open museum drawers to study familiar songbirds that had been collected from their faraway wintering grounds. Rows of brilliant orange-and-black Baltimore orioles bore tags naming foreign localities such as Guerrero, Mexico; Gamboa, Panama; and Copan, Honduras. What had the oriole been doing, and from what tree had it fallen after it was shot? Which voices of tropical birds had resumed after the momentary silence following the crack of the gun? How had the oriole's life differed on its wintering grounds, surrounded by plants and wildlife unique to the tropics? In recent decades, more ornithologists have been traveling to the tropics to seek answers to their own questions, building a more complete picture about the lives of migrants in places where the birds were spending more than half the year.

For the most part, migrants in the tropics inhabit places that resemble their summering areas, but with striking differences in details. Bicknell's thrushes skulk in dense tangles of vegetation on cloud-shrouded mountains — whether in stunted forests of fir or among the giant *palo de vientes*. Baltimore orioles often inhabit open areas with trees, whether building their nests in the cottonwoods of Colorado or probing flowers of inga trees for nectar on coffee farms in Chiapas, Mexico. Whether in a hardwood swamp in the northeastern United States or a wooded park in Venezuela, Louisiana waterthrushes walk along streams, bobbing their tails and tossing leaves aside with their bills in search of insect larvae.

In comparison with the leafless wintry places the birds left behind in the north, the tropics offer a greater bounty of insects, fruit, and nectar. When the migrants arrive, however, they join an assemblage of resident tropical species, doubling the number of birds in some habitats. They also arrive during the dry season, the time of year when insect abundance in the tropics is thought to be at its lowest. On Guatemalan coffee plantations where 70 percent of the birds may be Neotropical migrants, the birds make a measurable dent in the abundance of insect prey. By constructing exclosures to keep birds away from patches of plants, a team led by researcher Russell Greenberg found that birds were reducing the number of large arthropods by 64 to 80 percent. This result suggested that birds were probably competing for limited food.

As migrants forage in the tropics, they sometimes exploit foods radically different from their summer diet. The same Cape May warblers that search the leaves for spruce budworms in summer spend tropical winter days piercing the base of flowers with their thin bills, then sipping the nectar. White-eyed vireos consume all kinds of prey on their breeding grounds—including butterflies, cockroaches, stink bugs, wasps, spider egg cases, and snails—as well as the berries of poison ivy and other plants. On Mexico's Yucatán Peninsula, they specialize in dining on the tiny berries of gumbo limbo trees, squeezing open the outer greenish capsules with their bills and consuming the red fruit inside. Only the seeds regurgitated by birds will sprout, so the vireos help generate the forest and sow their own crops.

Some birds partake of foods that don't even exist on their breeding grounds. In the mountains of Mexico, warblers flutter up against the blackened trunks of oak trees, picking off glistening droplets that appear to be coming from white hairs in the bark. The hairs actually belong to an insect that lives in the bark, drinks sap, and extrudes the droplet from the hair attached to its back end. Yellow-rumped warblers often monopolize the nectar, but when they have had their fill, Townsend's warblers and Wilson's warblers take their turn at the sweet straws. The remaining dew falls onto the trunks and nourishes the black fungus that covers the bark.

On Barro Colorado Island in Panama, thrushes, tanagers, and warblers join resident antthrushes, antbirds, and woodcreepers that follow swarms of army ants. Wood thrushes often hop on the forest floor, snapping up insects as they flee from the advancing swarm. Passing through on migration, Canada warblers dance along the low vegetation above the ants, sallying out to catch the insects that take to the air. Summer tanagers flitting in the trees above snap up escapees missed by the warblers.

The foods that birds exploit on their wintering grounds also influence their social lives. During the breeding season, most migratory songbirds are territorial — a mated pair defends an area around their nest that contains the food they need for themselves and their young. In the tropics, some migrants roam widely in flocks; others pass a solitary winter in the same patch of forest. Still others may be solitary or sociable depending on the circumstance or the time of season.

Birds often travel in flocks if their favorite food is difficult to monopolize — either because it occurs in ephemeral patches or because it is so abundant that for a while there is plenty for all. In their brown winter plumage, hundreds or thousands of indigo buntings roam the fields by day, feeding on seeds and buds. At night, they roost together, in contrast with their solitary sleeping habits in summer. Swainson's thrushes also travel in flocks, but they move in search of ripening fruit rather than grain. They also roost together at night. Bay-breasted warblers defend territories when they first arrive on their wintering grounds, but when fruit and nectar become more abundant later in the season, they forage in flocks.

Some birds are sedentary all winter, but join in with mixed flocks of migrants that pass through. Usually only one or a few of each species join in, each finding food according to habit. A flock in Cuba might include Cuban vireos, Cuban todies, loggerhead flycatchers, and an assortment of migratory warblers. Worm-eating warblers pry open dead curled leaves, then gobble up the bonanza — ten times more arthropod prey than on a similar-sized green leaf. Black-and-white warblers inch along tree trunks, stabbing at the bark with their bills, and black-throated blue warblers glean insects from the lush foliage. In this way, the birds interfere little with one another's food finding, but gain the group advantage — more eyes and ears on the alert for predators such as Cuban boas and Gundlach's hawks.

For a long time, it was a common assumption that most migrant songbirds roamed from place to place in search of food during winter. In the 1950s, however, Paul Schwartz, an American living in Caracas, Venezuela, noticed that the American redstarts in the wooded ravine around his home seemed to remain there for the entire winter. He often sat on his porch watching a redstart as it flitted about in the vines so near that he could have touched it. Over the years, he began to suspect that the same birds were coming back year after year. If there was a female in one year, a female appeared the next. One winter, the redstart inhabiting his porch was a young male with grayish head and breast and olive wings and tail, splashed with patches of yellow. The next winter, a male redstart appeared — charcoal black with streaks of orange, and a peculiar mix of colors on its breast. Schwartz suspected it was the same male, transformed into adult plumage. The next winter, the redstart reappeared, recognizable by the pattern on its breast.

Curious to know more, Schwartz decided to mark migrants with unique combinations of colored plastic bands. He focused on banding northern waterthrushes because of their abundance in a nearby botanical garden. Over the years, Schwartz confirmed his hunch that the waterthrushes were defending territories, each bird for his or her own. A waterthrush patrolled its area by walking on the ground or hopping from perch to perch, uttering *tink, tink* to announce its presence. If it encountered an opponent, it would approach in a crouched walk with tail spread and wings quivering. If the challenger responded likewise, the birds would edge closer to each other, then fight in a fury of blows from wing and bill. The winner remained on its territory from at least October well into April, twice as long as its residence on territory in summer. Waterthrushes were so fixated on their winter homes that when Schwartz captured four and drove them to new locations as far as forty miles away, they all returned within twelve days.

During the 1970s, John Rappole and Dwain Warner documented that winter territoriality was common among migrants. In the Tuxtla Mountains of southern Veracruz, Mexico, fourteen species defended territories in winter and returned in subsequent years, including wood thrushes, ovenbirds, yellow warblers, Wilson's warblers, magnolia warblers, hooded warblers, and American redstarts. Rappole and Warner also banded many "floaters," birds that presumably could not win territories in the limited forest available.

They also discovered a peculiar distribution of hooded warblers. In the rainforest, nearly all were mature males, with bold black hoods and bright yellow faces. In nearby areas of secondary forest, with younger trees, most of the warblers had the partial hoods and light yellow faces of females and yearlings. In 1982, a team led by James Lynch found that hooded warblers on the Yucatán Peninsula were also sexually segregated — the males in mature forests, and the females and young birds in scrubby abandoned fields. The few females that staked out territories in mature forests had relatively bold facial markings, resembling those of males.

Were male hooded warblers monopolizing the best areas, forcing females into second-class habitats? A team led by Lynch and Eugene Morton tested this idea in Quintana Roo, Mexico. The hooded warblers there lived in forests interspersed with shallow depressions, called tintales, with shrubby vegetation and small trees. Male hooded warblers staked out territories in the forested areas and females lived in the tintales. If the males were excluding the females from the forest, the females should move in, given the opportunity. However, when the researchers removed some of the males, females on adjacent territories stayed put, suggesting they were there by choice. Perhaps males were somehow better suited to life in the forests and females were better at finding food in open areas — a notion later supported by

experiments showing that females in captivity preferred artificial habitats with horizontal lines on the backdrop, sim-
ulating a shrubby environment. Then, in the 1990s, another
researcher studied a similar phenomenon among American
redstarts and came to a dramatically different conclusion,
one that would change the way people thought about the in-
timate linkages between events on the wintering and breeding grounds.

IN JAMAICA, visiting bird watchers are often surprised to
find American redstarts—one of the most beautiful warblers of North American forests—flitting about conspicu-
ously in garbage dumps. Seemingly oblivious of the stench
and the stomach-churning piles of waste, the redstarts appear as dapper as ever in their black-and-orange plumage,
twisting and turning in the air as they snap up flies attracted
to the garbage. Since the redstarts perform the same ma-
neuvers around outhouses, Jamaicans sometimes call them
"latrine birds." The nickname hints at the tiny warblers'
adaptability; they spend the winter in Jamaica wherever
they can, in lush mangrove forests, dry thorny scrub, shady
coffee plantations, smelly dumps, or gardens in downtown
Kingston. Since few of Jamaica's natural habitats remain,
this might be construed as good news for the future of
redstarts—unless there was a reason to think that all places
might not be equal.

On the breeding grounds, biologists had shown that the
quality of a bird's territory can influence how many young a
bird raises, and even affect its chances of surviving from one
year to the next. Yet until the 1980s, no one took a careful
look at whether a migrant's experiences on its tropical

wintering grounds might play a similar role. After seventeen years of studying warblers and other songbirds in New Hampshire, Dartmouth College biologist Dick Holmes and his research team traveled to Jamaica to find out how the birds' six-month stay on the island factored into the equation.

On the southwestern side of the island, Holmes and his team found Neotropical warblers in abundance — more than thirty species, including black-throated blue warblers, northern parulas, black-and-white warblers, northern waterthrushes, magnolia warblers, worm-eating warblers, and ovenbirds. Graduate student Peter Marra, who had studied redstarts on their breeding grounds, was particularly intrigued to find American redstarts in two habitats that seemed worlds apart from the hardwood forests of New Hampshire — and from one another, despite the fact that they were side by side.

The first was a black mangrove forest, flooded with swampy water and shaded by the dense green canopy forty feet above. Tiny vervain hummingbirds zipped from flower to flower in flashes of green. Bright green Jamaican todies darted among the leaves, gleaning insects. Caribbean doves walked on the mat of thin intertwined mangrove roots at the base of the trees. Northern parulas, magnolia warblers, and ovenbirds flitted in the trees, and American redstarts dived acrobatically, flushing insects from leaves by fanning their tails and drooping their wings. The active birds called frequently with a metallic chipping note, fighting for a place in the mangrove forest in spectacular bouts of circle chasing, flying back and forth at one another from perch to perch in great arcs, and sometimes grappling with one another in the air.

Where the mangrove forest ended, a thick tangle of thorny scrub began, choked with vines and interspersed with spindly logwood and poisonwood trees. Here, where the mangrove trees had been cut down for charcoal and fence posts, the sweltering sun beat down on the scrub and the bare limestone rocks underfoot. As the dry season advanced, the parched trees lost their leaves and many of the insects they had earlier harbored. The redstarts were here, too, but Marra had to look and listen for them more carefully. They were quieter, more reclusive, and they blended in better with their surroundings, since most of them wore the muted plumage of females and yearlings.

Indeed, Marra found that 70 percent of redstarts in the scrub were females, compared with only 30 percent in mangroves. Were most of the females living in the dry impoverished scrub by choice or because the males in the mangroves were keeping them out? Marra, Holmes, and collaborator Thomas Sherry put unique combinations of plastic colored bands on the birds' legs so they could recognize each warbler. In experiments similar to the ones that Lynch and Morton had tried with hooded warblers, they removed some of the birds and waited to see what happened.

One day later, new redstarts had moved onto 67 percent of the vacated territories in the mangroves, compared with only 8 percent in the scrub. A week later, all the vacancies in the forest were filled, compared with only half in the scrub. Furthermore, the proportion of females increased in both habitats, especially among the mangroves. The researchers concluded that a spot among the mangroves was desirable, and those redstarts in the scrub were forced out by more dominant birds. Marra also found that male redstarts were larger than females, and that females in mangroves were larger than those in scrub habitats — probably explaining why they had an edge in winning territories.

Later, as a scientist for the Smithsonian Environmental Research Center, Marra and collaborator Rebecca Holberton also showed that redstarts in scrub paid a physiological cost for living there. They collected blood samples from the redstarts and measured the concentrations of corticosterones in the blood. Corticosterones are known as stress hormones because they are released into the bloodstream in response to emergencies, such as when a bird is captured by a predator or has trouble finding enough to eat.

In October, soon after the redstarts had arrived in Jamaica, Marra and Holberton found that the birds in mangrove forests and scrub had similar concentrations of corticosterones in their blood. At that time of year, the wet season, both habitats were moist and probably hosted a good supply of insects. In March and April, as the pools of water in scrub habitats dried up, stress hormones shot way up for redstarts in the scrub. They also lost more weight than the birds in mangrove forests. Since the birds had no fat, and high levels of corticosterones are associated with muscle loss, the birds were probably surviving by burning up their muscle tissues. That meant that redstarts from the scrub were probably departing for the breeding grounds on a deficit — a dangerous situation given that some songbirds are fifteen times more likely to die during migration than at any other time of year.

The consequences were sobering. Based on the number of redstarts that returned to the study sites, Marra and Holmes found that the birds in scrub had shorter lives. The oldest redstart was banded as an adult and returned to the mangroves for eight consecutive years. In contrast, the longest-lived bird in the scrub was a female who returned only four years in a row. On average, the redstarts in mangroves lived about eighteen months, compared with a mere six months for redstarts in scrub. Additionally, the effects from a hard winter in the scrub appear to influence a redstart's chances of raising young the following summer.

Marra, collaborator Keith Hobson, and Holmes looked at carbon isotopes in the blood of redstarts arriving on the breeding grounds to determine what type of habitat they had come from. They found that males from wetter wintering areas arrived on their breeding grounds at the Hubbard Brook Experimental Forest in New Hampshire earlier than males that had been in drier habitats. Additionally, later arriving males were in poorer condition.

By arriving earlier, males from wetter areas could reap some important advantages. They would have the best pick of territories and could begin mating sooner. They would have greater chances of raising more young, since clutches laid earlier in the season tend to have more eggs. Additionally, they would have more time to renest in case of failure and more time to feed fledglings while the woods were still thick with insect prey.

A team led by Marra and graduate student Ryan Norris found similar results in Chaffey's Lock, Ontario — males from better wintering habitats arrived earlier. As a result, males from good wintering areas could be expected to raise one more fledgling than males from poorer areas. Although wintering habitat did not influence the arrival date of females, it did influence when they laid eggs. Because females from the best habitats laid eggs sooner, they reaped the same benefits earned by early arriving males: larger broods that fledged earlier. The team estimated that females from the best wintering areas could produce two additional offspring and raise them to fledging almost a month sooner than females from poor wintering areas. These results quantified for the first time how the quality of a migratory bird's winter residence could influence the number of young it would send forth the following summer.

If segregation of the sexes on the wintering grounds is any indication, many songbirds face the same challenges that American redstarts do. For at least fourteen Neotropical migrant species, including common yellowthroats, magnolia warblers, and northern parulas, scientists have documented males and females using different habitats on their wintering grounds. In most cases, the brightly colored males predominate in relatively mature forests, while the subtly plumaged females and yearlings inhabit disturbed areas, including regenerating forests, pastures, and cornfields.

In Ecuador, the pattern follows the slopes of mountains: from low to high elevations, the sex ratio for summer tanagers switches from mostly females to mostly males. There is another geographic difference, too. Male summer tanagers tend to winter in more northerly areas; the females travel farther south. The skewed distribution of male and female migrants elsewhere also hints that females are sometimes forced to travel farther to find any suitable habitat at all. For example, nearly all American redstarts and black-throated blue warblers are females on San Salvador, a dry and isolated island in the Bahamas.

Findings such as these testify to the limited availability of good habitat in the tropics. Just because the "latrine bird" can survive around garbage dumps and in inhospitable scrub doesn't mean that the redstarts can do without the mangrove forests. Not much native forest remains in the lowlands of Jamaica, and one of the forests where Marra studies the redstarts has been proposed as a new site for a refinery and an airstrip. If that happens, more birds would be forced out into the scrub, displacing females and young birds from even that hardscrabble place. More redstarts in the scrub, in dumps, and downtown Kingston might create the impression that these birds are opportunistic and adaptable when another way to look at it is they could be desperate.

In spring in Maryland, near where Marra lives for most of the year, the first American redstarts usually turn up around April 20. The first spring arrival is usually a handsome male, perched up in a small tree or shrub, repeating his high-pitched song, *Tsee tsee tsee TSEE-o. Tsee tsee tsee TSEE-o.* If the studies are right, he probably spent the winter in a shaded forest—perhaps a mangrove forest or a shade coffee plantation. Two weeks later, a female redstart appears, flitting in the young leaves of an oak tree, flashing her wings and tail to startle insects out of hiding. Chances are that she inhabited a harsher place and found less to eat during her last days in the tropics. By late May, when most of the redstarts have paired up, the newly arriving birds slow to a trickle, including those that endured such a difficult winter that they almost didn't make it back.

After witnessing the tussles and stresses of life in the tropics, Marra can't help but to think about the Caribbean when he sees an American redstart in Maryland. For the average person, in contrast, the connection is remote. Birds always return in spring, singing and courting, seemingly unweathered by the many miles we know they must have traveled. Beloved as redstarts are, one looks so much like another that it's hard to think very much about their individual fates. That's one reason why Marra isn't simply content to study redstarts. He has looked for a way to bring the powerful message of bird migration home to anyone who has a migratory songbird in his or her backyard.

In 2000, Marra started up Neighborhood Nestwatch in Maryland, a program that takes him and other scientists to people's yards to capture and band birds. Participants can watch as Marra sets up mist nets outside their houses and then removes birds that have flown into the nets. After putting colored leg bands on a gray catbird and taking its measurements, Marra hands the dusky bird with a black cap to participants to hold. Cupping the catbird in their hands, they can look into its alert dark eyes and feel beneath the soft feathers its rapidly beating heart. Then they release the bird and keep watch over it during the summer. They look for nests, check for the bird's bands when it appears in their yard, and write down when they see the bird and when they don't.

Marra uses their observations to learn how wild birds are coping in urban areas — a subject about which scientists know surprisingly little. At the same time, he has found that it's a great way to get participants to think about the same things he does: the great mysteries of migratory birds.

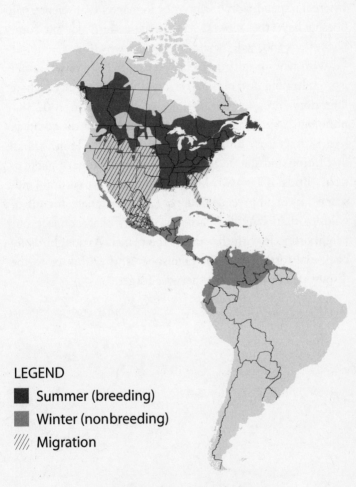

LEGEND

■ Summer (breeding)
■ Winter (nonbreeding)
/// Migration

American redstart migration map.

People who watch a banded gray catbird outside their window all summer will find it hard not to wonder exactly where it's spending the winter, or to marvel that science still doesn't have the answer. And if the catbird doesn't come back, they, too, will inevitably wonder why.

With new insights gained in recent years, conservationists are leading international efforts to save the places where migratory birds live at all times of the year. In 2002, ornithologists exchanged scientific findings about the complex lives of migratory birds in a symposium at the Smithsonian Institution and shared their findings in a landmark publication, "Birds of Two Worlds." As humans alter natural landscapes at an unprecedented pace, it's fortunate for all of wildlife that songbirds are messengers that connect our neighborhoods with the rest of the planet. As songbirds fly back and forth across the hemispheres, they link us to the distant places we might otherwise forget.

11
*B*IRDS OF NORTHERN WINTERS

Ere long, amid the cold powdery snow, as it were a fruit of the season, will come twittering a flock of delicate crimson-tinged birds, lesser redpolls, to sport and feed on the seeds and buds now just ripe for them on the sunny side of a wood, shaking down the powdery snow there in their cheerful social feeding, as if it were high mid-summer to them.
— Henry David Thoreau

IN THE HEART OF WINTER in Concord, Massachusetts, Henry David Thoreau explored the snowy woods and fields, looking for signs of birdlife. He climbed snowbanks to search among bare branches for abandoned nests, each one filled with a globe of snow—an "ice-egg," he called it. He wrote about nests of silver birch and thistledown that had once held warblers, nests from which birds had radiated and followed the sun. Yet even after most birds had long since departed for warmer places, Thoreau still heard the cries of blue jays; to his ears they were like frozen music of the winter sky. On a snowy pond, he found bird tracks leading from ice hole to ice hole, where crows had looked for fish bait left by fishermen. High above whitened fields, he watched a flock of two thousand snow buntings, like snowflakes against the clouds.

The songbirds that had visited in summertime — vireos, thrushes, catbirds, swallows, warblers, tanagers, orioles, grosbeaks, and buntings — had all flown south, heeding the signal of shortening autumn days. Through even the coldest months, though, the crows, jays, nuthatches, and chickadees remained, subsisting on winter's fare, which they shared with new visitors from farther north — snow buntings and tree sparrows from the arctic tundra, and crossbills, pine grosbeaks, and common redpolls from the boreal forests. Throughout the winter, Thoreau observed the changes in birdlife around Walden Pond, a single place that hinted at the complex and varied nature of migration across the continent.

Within the United States and Canada, the winter distributions of birds are dynamic and complex, reflecting differing strategies for surviving times of cold and scarcity. Some bird species are resident year-round, but others wander nomadically, continually seeking food in new areas. Some species migrate every year, spending summer and winter in completely different areas throughout the species' range; others undertake such migrations only in years when they cannot find enough food, or are resident in some regions but migratory elsewhere. Because of this variation, and the difficulty of tracking migration across wide areas, the movements of even our most familiar winter birds are poorly understood. Recently, however, researchers have used observations collected by birders in thousands of different locations, revealing how changes in any single place fit into an ever-changing panorama of bird migration that no one person alone can see.

THOUGH THE ONSET of cold weather often coincides with migration, many birds can survive frigid temperatures as long as they have enough food to eat. Thoreau observed flocks of lively common redpolls in bare birch trees during the chill of December, a sight as surprising to him as though he had found brilliant crimson flowers flourishing in the snow. He wrote in his journal that the maker of redpolls "made bitter imprisoning cold before which man quails, but He made at the same time these warm and glowing creatures to twitter and be at home in it."[1]

Common redpolls live in some of the coldest places in the world—the boreal forests and taiga of Canada, Alaska, Scandinavia, Russia, Greenland, and Iceland—sometimes remaining through winter, and can survive temperatures that plummet to sixty-five degrees below zero. (Their relatives, hoary redpolls, can withstand even colder temperatures, to eighty-eight degrees below zero.) In a world where water is often bound up in snow and ice, common redpolls bathe in the snow, fluttering their feathers in the cold powder and leaving pock marks in the snow when they are done. Before dark, they sometimes burrow into the snow with their feet and bills, excavating tunnels up to sixteen inches long.

There, they spend the night in a snug chamber at the end of the tunnel, then break through the snowy roof to begin the new day.

Black-capped chickadees survive extremely cold nights by roosting in tree cavities, reducing radiant heat loss by 60 to 100 percent compared with roosting in open air. They can also lower their body temperature overnight by as much as twenty-one degrees, sleeping in a state of hypothermia that conserves energy. During the short days of winter near Fairbanks, Alaska, chickadees must eat enough food during 3.5 hours of daylight so that they can fast for the next 21.5 hours in the chilly darkness.

To survive cold temperatures, birds face a huge challenge: they must be able to find enough food to generate continuous body warmth. Common redpolls eat as much as 42 percent of their body weight in seeds during a single day, the equivalent of sixty-three pounds of food for a 150-pound person. While looking for food in exposed trees and shrubs, a redpoll can quickly swallow seeds and hold them in expandable portions of its esophagus called the diverticula. The diverticula can carry 15 percent of the redpoll's body weight in seeds. For a 150-pound person, that would be analogous to carrying twenty-two pounds of food in one's throat. After filling up with seeds, the redpoll can retreat to a sheltered spot and regurgitate, husk, and consume the seeds at leisure.

Red crossbills can store three hundred spruce seeds in a portion of their esophagus called the crop. Before they roost for the night, they fill their crops with seeds—midnight snacks, perhaps, to fuel them through until morning. Their daytime consumption of seeds is probably formidable; their relatives, white-winged crossbills, must eat about one seed every seven seconds during a wintry eight-hour day to survive. Thoreau observed a pair of red crossbills holding hemlock cones in one foot and extracting seeds with their bills so quickly, one cone after another, that he couldn't see the crossed bills in the blur of constant motion.

The upper and lower bills, which cross at the tip, allow crossbills to twist the stalks of coniferous cones, then bite between the scales to expose the seeds, a specialized skill that enables them to exploit food inaccessible to other birds. The cones of different tree species vary in size, and crossbills in different regions have bills that are suitably sized for opening the predominant cones they encounter. Red crossbills with small bills are best at opening small cones of hemlocks, whereas crossbills with larger bills can extract seeds more effectively from cones of ponderosa or lodgepole pines.

In contrast with the extreme specialization of crossbills, omnivorous American crows survive winters by eating whatever they can find. In summer, their feasts include toads and fish, earthworms and insects, corn and oats, bird eggs and nestlings, and wild cherries. In winter, Thoreau watched them dining on acorns. In frozen droppings left by crow tracks on the icy pond, he found indigestible bits of red cedar berries and barberries. Once a friend brought him the contents of a crow's stomach in a jar—a winter's diet of apple pulp and skin, pieces of skunk-cabbage berries, and small bones, perhaps from a mouse.

Some birds have specialized digestive systems that enable them to eat winter foods that most other birds pass up. Yellow-rumped warblers migrate south from the coniferous forests of Alaska and Canada where they breed, but can spend their winters farther north than most other warblers because they can eat wax myrtle berries and bayberries. These fatty berries contain wax (colonial settlers used them to make candles), a high source of energy that other warblers can't digest. Yellow-rumped warblers have high concentrations of bile salts in their gallbladders and intestines and can move partially digested berries from their intestines into the gizzard to aid in digestion. Because of this ability, yellow-rumped warblers can winter wherever they find wax myrtle berries and bayberries, including the Atlantic Coast north to Massachusetts, and inland on the Canadian shore of Lake Erie. Tree swallows, the only North American swallows that eat fruit, can also digest wax myrtle berries and bayberries. They may spend the winter in coastal locations as far north as Long Island, New York, well beyond the northern limits of bank swallows and cliff swallows, relatives that travel all the way to South America.

Winter weather is deadly when it prevents birds from finding food. Carolina wrens usually remain on the same territory with their mates all year long, searching for food on the ground among plants and fallen leaves. Warming temperatures since the late nineteenth century have allowed Carolina wren populations to expand northward into Michigan, Massachusetts, and New York, but severe winters can lock the ground in snow and ice. When that happens, many Carolina wrens cannot find enough food and die of starvation or exposure. Data from participants in the Christmas Bird Count show that Carolina wrens declined by 50 percent from 1977 to 1978, after a winter with below-average temperatures. In subsequent years, numbers climbed again, but in 2003, abnormally cold temperatures and high snowfall were disastrous for Carolina wrens in some areas. In 2004, participants in the Great Backyard Bird Count reported Carolina wrens on 56 percent fewer checklists than the previous year at the northern edge of the species' range in Connecticut, Massachusetts, New York, and New Hampshire. In extreme winters, populations may shrink away from the coldest areas; then in milder years, they build up, and new generations push the limits of the northern boundaries.

As insurance against hardship, some birds store food for weeks or months. Black-capped chickadees hide seeds in bark, pine needles, dirt, or snow. They can recall these locations and retrieve food for at least twenty-eight days afterward. They even remember where they have stored the highest-energy items. These abilities require specialized brains; researchers have found that the hippocampus, the portion of the brain involved in spatial memory, is proportionately larger in chickadees than in birds that don't store food.

Phenomenal spatial memories also help jays and nutcrackers to survive periods of scarcity in harsh climates. Blue jays may hide several thousand acorns and nuts in autumn, burying each one in the ground and covering it with a dead leaf or pebble. Gray jays, permanent residents of boreal and subalpine forests, store food in trees above the snowline. First they coat the food with sticky saliva, then attach the items beneath bark, under lichen, or among leaves. On a summer's day in Alaska, they may store food in one thousand locations in a seventeen-hour day, more than one item per minute.

In montane forests, a Clark's nutcracker stores as many

as 98,000 pine seeds in the ground or in trees and remembers the locations for as long as nine months afterward, even when covered by snow. A flock of 250 pinyon jays can put away 4.5 million pine seeds in a single autumn, when seeds are plentiful. Because the pinecones often occur in clumps at scattered locations, the pinyon jays may harvest the seeds, cram them into their throat pouches, and travel as far as seven miles back to home base, where they hide the seeds in the ground.

WHETHER OR NOT they store food, birds must be prepared to move when food runs out, or suffer the fate of Carolina wrens. The availability of food is a major factor underlying the migrations of birds. Every year, snow buntings and American tree sparrows undergo regular long-distance annual migrations, similar to those of some Neotropical migrants, except that their movements span northerly places such as the tundra of Greenland and the snowy fields of Massachusetts. Snow buntings breed in rocky habitats on the high arctic tundra, tucking their nests into cavities in rocks, under boulders, in cracks on cliffs, or—if no suitable place is found—in skulls or discarded tin cans. Their nestlings grow in a cup of moss lined with the fur of arctic foxes, lemmings, and caribou and the feathers of snowy owls. In September, these "arctic snowbirds," as Thoreau called them, leave behind their summering grounds before they become unforgivingly cold, dark, and devoid of food. "Like a snowstorm, they come rushing down from the north," Thoreau wrote. "If New Hampshire and Maine are covered deeply with snow, they scale down to Massachusetts for theirs. Not liking the grain in this field, away they dash to another distant one, attracted by the weeds rising above the snow . . . Not the fisher nor the skater range the meadow a thousandth part so much in a week as these birds in a day."[2] Tree sparrows, birds that nest in tussocks of grass on the arctic tundra, also come south in winter, traveling hundreds or thousands of miles to look for dry winter weeds that they can harvest for seeds. Thoreau watched as the sparrows flew up from the ground to beat at the weeds with their wings, scattering bountiful seeds on the snow, he said, as though they were shaking down crusts of bread to be eaten from a white tablecloth.

In contrast with the reliable appearance of snow buntings and tree swallows, Thoreau noted, common redpolls could be the most abundant songbird in Concord one winter, completely absent the next year, and abundant again the year after that. Redpolls, along with pine grosbeaks, purple finches, pine siskins, and evening grosbeaks, are known as "winter finches" because they are seen south of the boreal forests some winters when there are seed shortages. In response to food shortages, several other species—including Clark's nutcrackers, pinyon jays, red-breasted nuthatches, red crossbills, black-capped chickadees, and boreal chickadees—also travel beyond their normal range in massive movements called irruptions.

Numerous observers besides Thoreau have noted throughout the decades that some bird species undergo cyclical irruptions. Using ten years of data from participants in the Christmas Bird Count, scientists Carl Bock and Larry Lepthien published a study in 1976 that confirmed widespread biennial irruptions among several seed-eating birds, including common redpolls, black-capped chickadees, and red-breasted nuthatches. These species were relatively scarce in years that began with even-numbered winters but appeared south of their breeding ranges in odd-numbered winters. Seed crops in boreal forests throughout North America fluctuate in patterns of abundance and shortfall from one year to the next, driving the on-again, off-again presence of boreal seed-eating birds across the northern United States and southern Canada. In good years, the birds stay farther north and produce many young, boosted by the abundance of food. If the trees produce few seeds the next year, especially high numbers of birds are confronted with a diminished food supply, causing them to move south the next autumn.

Using data collected by participants in Project Feeder-Watch, a winter-long survey of feeder birds, scientists found that common redpolls moved farther during some irruptions than others, suggesting that the birds stopped when they found sufficient food. Banding records of common redpolls show that even so, some redpolls travel immense distances. Redpolls banded in the northeastern United States were recovered as far away as Alaska and the Khabarovsk Krai, Russia. Bock and Lepthien speculated that mortality of redpolls is probably very high during irruptions and that boreal trees could benefit from synchronized seed production every other year, resulting in reductions of birds and other seed predators.

Because different bird species eat seeds produced by different kinds of trees—including birch, alder, maple, and conifers—researchers believe that different species of trees in the boreal forest may undergo synchronous cycles of seed production. Recent studies using thirty-eight years of data from the Christmas Bird Count show that red-breasted nuthatches, pine grosbeaks, common redpolls, pine siskins, and evening grosbeaks tend to irrupt synchronously across North America.

However, patterns of movement across wide geographic areas and long time scales are more complicated. Over the long-term, Christmas Bird Count data showed that certain species, such as red-breasted nuthatches, pine siskins, and common redpolls, irrupted synchronously during one decade, but not in another. Some groups of seed-eating birds tend to have synchronized movements only in certain regions, such as purple finches, evening grosbeaks, and pine siskins in the West. During a ten-year period between 1987 and 1997, counts from Project FeederWatch participants showed that pine siskins irrupted on opposite sides of the continent in alternate years.

Unpredictable irruptions of other seed eaters may be related to irregular factors that cause seed crop failures. During the winter of 2002, Project FeederWatch participants in some areas reported unusually high numbers of pinyon jays, Steller's jays, and Clark's nutcrackers at their feeders. Fires, drought, and bark beetle infestations may have caused seed crop failures that forced the birds out of their usual areas.

Some birds are both irruptive and nomadic — they roam freely throughout the winter, but travel even farther than usual when they can't find enough food. Fruit-eating bohemian waxwings sometimes appear in areas well to the south of their usual range, as they did in 1959, when more than ten thousand passed through Santa Fe, New Mexico. Nomadic red crossbills search continually for coniferous cone crops in taiga and montane forests. Remarkably opportunistic, they will even begin nesting as early as December if they happen upon a good supply of seeds. When their food supplies fail, they skip breeding altogether as they wander in search of food.

Other songbirds migrate more predictably, but not everywhere throughout the species' range. These birds are called "partial migrants" because only a part of the population moves while others remain resident. When the first snow covers the seed-bearing weeds, American goldfinches leave the north in flocks of hundreds and join resident goldfinches in fields and at feeders farther south. Winter roosts of hundreds of raucous crows in the Northeast may include birds from Canada as well as those from local areas. Dark-eyed juncos move out of most of Canada in winter, but in the Appalachian Mountains, male juncos may remain in the same area all their lives or move to lower elevations for the winter. Females and young birds often travel farther than adult males, whether migrating down mountainsides or moving from north to south.

Scientists hope to learn more about year-to-year variations in the movements of millions of birds that remain north of Mexico by analyzing data that thousands of birders contribute to citizen-science projects. These require varying levels of commitment, but many of them are simple enough even for beginning birders (appendix B). Some projects make the data available for all to see. For example, with eBird, an online checklist project from the Cornell Laboratory of Ornithology and Audubon, birders can record counts at any time of year, as often as they like. Once entered, the counts become part of a central database, accessible to birders, scientists, and anyone else with Internet access. Participants can look back at the numbers and kinds of birds they have seen at their favorite spots in different seasons and years and can explore what others are seeing in any region, through maps, graphs, and summaries online. By doing so, they may recognize some of the complex patterns of winter migration or may find new ones that scientists have not even noticed yet. By taking a few moments to send in observations, birders are turning their own winter bird sightings into pieces of a puzzle that no single person or team of ornithologists could assemble on their own, a lasting record that will be used for years to come in understanding the birds that spend their winters closest to our homes.

OF THE BIRDS that migrate within the arctic, temperate, and subtropical regions of North America, about 78 percent have some populations or closely related species that reside in the tropics year-round. Ornithologists believe that most of North America's birds originated in the tropics long ago and that, through time, birds gradually ventured northward, encountering both challenges and opportunities along the way. Some traveled to places where they feasted on insects and other foods abundant in summer, then retreated back south in winter. Ultimately they traveled farther and farther to find good unoccupied places to breed, evolving the ability to migrate on tightly timed schedules, traveling thousands of miles by instinct. As these birds flew south out of the shadow of winter and returned with the lengthening days of spring, other birds adapted to the northern landscape's snowy climate and seasons of scarcity.

Upon seeing flocks of redpolls feasting on birch seed in December, Thoreau wrote, "These crimson aerial creatures have wings which would bear them quickly to the regions of summer, but there is all the summer they want."[3]

A life in the north comes with a price, though. The birds that reside in cold wintry places are likely to have even shorter lives than the migrants that travel thousands of miles back and forth each year. To beat the odds, they rely on their innovations — their snowy excavations, their hypothermic slumbers, their deft harvest of seeds from pinecones, their astute memories for recovering stored food, their tricks for digesting waxy substances useless to other birds. They take the gamble that they will survive, that they will be able to choose a place to breed perhaps even before the first Neotropical migrant crosses the Gulf of Mexico, and that they will soon harvest the gifts of summer for their young.

In the far North, crossbills roam the ice-clad forests and redpolls leave the wild boreal forests, surging southward and flocking at backyard feeders. American robins roam in search of berries, often keeping south of the snowline, where they can also probe for invertebrates in the soil. Yellow-rumped warblers descend on wax myrtle bushes on the shores of the frozen Great Lakes while those wintering in the mountains of Mexico flutter near the trunks of trees to suck sweet juices from insects in the bark. In the Caribbean and Central and South America, warblers, vireos, tanagers, buntings, and other birds return to the land of their ancestors, joining resident tropical birds until spring comes again. At any given place, whether in the high arctic tundra, the snowy fields of Massachusetts, or tropical forests, the composition of birds changes from season to season and year to year. As they have for millennia, the birds continue their seasonal journeys over the surface of the earth, following the flush of plants and insects stirred to life as the planet spins on its tilted axis around the sun.

Hot Spots for Watching Birds in Winter, from Near to Far

There are many rewards to watching birds in your own neighborhood and local parks in winter, when you can get to know birds that inhabit your area year-round as well as those whose presence and numbers fluctuate with weather conditions and the abundance of natural food supplies. Repeated observations at the same locale over time are particularly valuable for scientific studies. In particular, eBird, a project of the Cornell Lab of Ornithology and Audubon, encourages birders to participate in the eBird Site Survey, which provides guidelines for how to turn bird sightings from your favorite locale into useful observations for understanding the distribution, abundance, and migratory patterns of birds. For details, visit http://www.ebird.org/content/news/essintro.html.

Winter is also a great time to travel farther afield, following Neotropical migrants south of Canada and the United States, where they join dazzling Neotropical resident species for much of the year. More than 80 percent of Neotropical migratory bird species travel through or winter in Mexico. The Yucatán Peninsula offers opportunities to see trans-Gulf migrants among familiar resident species that are found on both coasts, as well as birds endemic to Mexico. Colima and Jalisco on the Pacific Coast are also phenomenal places for seeing migratory and resident species. All Neotropical song-birds that breed west of the Rockies can be found there as wintering species or passage migrants. The Caribbean is a migratory route or wintering grounds for many species that breed in the eastern United States. The forests of Jamaica and the slopes of the Bahoruco Mountains in the Dominican Republic attract many Neotropical songbirds. In Central America, ecolodges are surrounded by fantastic natural areas with trails that are ideal for watching birds on your own or with the help of guides. Popular birding destinations include the Panama Canopy Tower in Panama, La Selva Biological Station in Costa Rica, and Chan Chich Lodge and La Milpa Field Station in Belize.

T'isil Nature Reserve and Celestún Biosphere Reserve, Yucatán, Mexico

More than 540 bird species have been recorded on Mexico's Yucatán Peninsula. Northwest of Cancún, the forests and trails of the privately owned T'isil Nature Reserve are accessible to guests of the Santa María Lodge. The reserve is rich with birdlife including eight oriole species, sixteen species of Neotropical migrant wood warblers, and regional endemic species such as the olive-throated parakeet, black-headed trogon, turquoise-browed motmot, and rose-throated tanager.

On the northwestern side of the Yucatán Peninsula, coastal dunes, mangrove forests, deciduous forests, and savannas offer a diversity of habitats for migratory and resident species. More than three hundred bird species have been recorded in the Celestún Biosphere Reserve, including endemic Mexican sheartails and Yucatán wrens. As many as thirty thousand greater flamingos can be seen by taking a boat trip to the estuary.

WHEN TO GO: Look for wintering migrants from October to May. Peak migration is early October in fall and late April to early May in spring.

NEAREST BIG CITY: The Celestún Biosphere Reserve is about fifty-six miles west of Mérida.

SPECIAL ACTIVITIES: The Yucatán Bird Festival is held in Mérida in November or December. The program includes lectures and field trips. For details, visit http://yucatanbirds.com.

NEARBY SONGBIRD HOT SPOT: Isla Cozumel is the largest island offshore of the Yucatán. Four endemic bird species— the Cozumel emerald, Cozumel wren, Cozumel thrasher, and Cozumel vireo—are found there.

FOR MORE INFORMATION: Santa María Lodge: (936) 321-0939 or (800) 583-6436; http://www.sitetraffic.com/sml.

RECOMMENDED READING: "Birds of Celestún Biosphere Reserve," by David Bacab. http://www.earthfoot.org/places/mx002.htm.

Soberanía National Park, Panama

At the Canopy Tower Ecolodge and Nature Observatory, visitors live in a tower surrounded by semideciduous rainforest in the Soberanía National Park. From the dining room on the top floor, panoramic windows look into the treetops shared by wintering migrants such as bay-breasted warblers and yellow-throated vireos, resident blue cotingas, chestnut-mandibled toucans, honeycreepers, slate-colored grosbeaks, blue-crowned manakins, and hundreds of other bird species. Along the trails in spring, scarlet tanagers, Swainson's thrushes, rose-breasted grosbeaks, vireos, and warblers can be found among trogons, motmots, puffbirds, manakins, antbirds, and other tropical species. More than 280 bird species have been observed from the tower and along a one-mile stretch of road nearby.

WHEN TO GO: December to March for wintering migrants and residents; mid-March to May for spring migration; September to the first week of November for fall migration.

NEAREST CITY: The drive from Panama City to the Canopy Tower is about forty-five minutes.

NEARBY SONGBIRD HOT SPOTS: Canopy Lodge at El Valle de Antón; Smithsonian Tropical Research Institute, Barro Colorado Island, Pipeline Road (see appendix A).

SPECIAL ACTIVITIES: For additional fees, the Canopy Tower offers daily tours to nearby birding spots, including Pipeline Road, one of the most famous birding spots in the world.

FOR MORE INFORMATION: Canopy Tower: (507) 264-5720; http://www.canopytower.com.

ℰPILOGUE

No Bobolink — reverse His Singing
When the only Tree
Ever He minded occupying
By the Farmer be — Clove to the Root —
His Spacious Future —
Best Horizon — gone —
Whose Music be His
Only Anodyne —
Brave Bobolink —
— Emily Dickinson

BOBOLINKS AND OTHER BIRDS inhabit landscapes that have changed dramatically since 1830, the year Emily Dickinson was born. The human population in the United States has skyrocketed to more than 281 million. Tractors have plowed nearly all of the native tallgrass prairie that once stretched from Canada to Texas.[1] East of the Great Plains, logging machines cut down nearly 400 million acres of virgin forest between 1860 and 1920. To-day, less than 1 percent remains of the virgin forest that once existed east of the Rockies. In Mexico and Central America, key wintering areas for Neotropical migrants, less than 2 percent of tropical deciduous forest may be left.[2] In Argentina, where bobolinks spend the winter, all but 10 percent of natural grasslands have been transformed into farms and cities.[3]

The list of vanished natural habitats goes on. Birds have fewer choices about where to live, raise their young, and seek shelter during migration than they did 150 years ago. The places that do remain for birds are often suboptimal. In the absence of tallgrass prairie where bobolinks once nested, hayfields beckon as safe places to hide nests in the grass, but mowers often destroy the eggs and young before they have a chance to fledge. A study in New York found that 80 percent of bobolink nestlings survived in undisturbed fields, compared with only 6 percent in fields that had been mowed, raked, and baled. The number of breeding bobolinks in the United States and Canada has dropped by about 75 percent in the last twenty-five years.[4] Henslow's sparrows, grasshopper sparrows, and meadowlarks also breed in hayfields and lose their nests to mowing.

Meanwhile, woodland songbirds settle in remnant patches of forest and suburban woodlots, places that harbor abnormally high numbers of predators, such as squirrels, raccoons, and jays that prey on bird eggs and nestlings. When researcher David Wilcove placed nests with commercially supplied quail eggs in forests of different size, he discovered that predators raided nearly all of the nests in small suburban woodlots, compared with only 2 percent in the

Great Smoky Mountains National Park.[5] Although these rates may overestimate the actual amount of nest predation, they indicate that small patches of forest could be "sinks" — places that drain the productivity of songbirds that invest in nesting there but fledge hardly any young.

Another hazard in open habitats resulting from forest clearing and agriculture is the brown-headed cowbird, a species that lays its eggs in other birds' nests. The victims may abandon their nests, or may be duped into feeding the cowbird's nestlings at the expense of their own young. Cowbirds are more likely to find songbird nests in small fragments of forest than in vast areas of intact forest. Research in Delaware showed that wood thrushes were ten times more likely to be parasitized by cowbirds in neighborhoods and small suburban woodlots than in a larger woodland nearby.[6] In fragmented forests in southern Illinois, the combined impact of cowbirds and predators was devastating: cowbirds laid eggs in from 50 to 100 percent of songbird nests and predators destroyed 70 to 99 percent of nests with eggs.[7]

In urban and rural areas, two other major threats to songbirds are cats and windows. Feral and domestic cats are estimated to kill hundreds of millions of birds each year and at least another hundred million die each year from collisions with windows.[8] Birds crash into windows during the day when they mistakenly fly toward the reflections of trees, shrubs, and sky. As they migrate at night, they may be drawn toward windows with lights shining from the other side. Especially in rain or fog, nocturnal migrants collide with buildings when they approach the lights or flutter in the light beams until they become exhausted. The Fatal

Light Awareness Program has documented more than 140 bird species that collided with buildings in Toronto, Ontario, alone.

Communications towers are also deadly for migrating birds. The U.S. Fish and Wildlife Service estimates that at least 4 million birds die in tower collisions each year. The grim tally includes many Neotropical songbirds, including ovenbirds, American redstarts, wood thrushes, black-throated blue warblers, bobolinks, and yellow warblers, as well as species with rapidly declining numbers, such as golden-winged warblers, cerulean warblers, Bachman's sparrows, and bobolinks.[9] The U.S. Fish and Wildlife Service has advised against placing towers in areas with high concentrations of migrants and recommends using white lights rather than solid red or pulsating lights, which seem to confuse birds the most.[10] Unless other ways are found to reduce tower strikes, the hazards will continue to increase. In the United States, at least seven thousand new towers for mobile telephones, television, radio, and other communications are being built every year, adding to the sixty thousand existing towers that are lit for human safety.[11]

Songbirds that navigate around this hazard-studded landscape and travel south of the United States border encounter another peril: the caged-bird trade. For the domestic trade in Mexico alone, more than one hundred thousand painted buntings were trapped from 1984 through 2000. In 2001 and 2002, after an international trade ban was lifted, Mexico also exported more than six thousand painted buntings to Europe and Asia.[12] More are sold illegally, a practice that is commonplace but difficult to document, according to Eduardo Iñigo-Elias, a conservation specialist at the Cornell Laboratory of Ornithology. These losses are especially sobering considering that the painted bunting is on the Partners in Flight WatchList as a species of conservation concern and has declined by 50 percent in the past twenty-five years.[13] Bobolinks are still sold as caged birds in Argentina, as they were when Olin Sewall Pettingill traveled there in the 1970s.[14] Untold numbers of other migratory songbirds are also trapped for the caged-bird trade in Mexico, Argentina, Cuba, and other countries.[15]

On both their breeding and wintering grounds, songbirds also face invisible, immeasurable threats from the use of pesticides. About 2.5 million tons of pesticides are used worldwide every year. In the United States, about 670 million birds are directly exposed to pesticides on croplands each year and at least 67 million are killed, according to conservative estimates by Cornell University scientist David Pimentel and colleagues.[16] These numbers do not include birds that may suffer indirect effects, such as reduced availability of insect prey or compromised health. One pesticide ingredient, acephate, may even hinder birds' ability to migrate, since it interferes with their ability to orient as they fly, according to the Smithsonian Migratory Bird Center.

The harmful effects of pollution on songbirds are very difficult to measure. In parts of Europe, acid rain has been implicated in the deaths of forests by damaging trees and in a changed soil chemistry that is less suitable for the invertebrates birds need to form normal eggshells. Researchers suspect that acid rain may help explain declines of wood thrushes and other woodland birds in forests of the eastern

United States. Global warming may disrupt the delicate timing of songbird migration and the availability of insects in spring. Warmer temperatures could also eliminate some plant communities where they thrive now, forcing birds to travel farther north or diminishing the amount of habitat they have to breed.

What is the cumulative toll on breeding birds from current threats? According to Partners in Flight, twenty-nine landbird species have declined by more than 50 percent since the late 1960s.[17] Radar images analyzed by Clemson University's Sidney Gauthreaux indicate that the volume of migration over the Gulf of Mexico during a three-year period in the 1980s had diminished by about half compared with three years during the 1960s.[18] As a group, grassland birds, including bobolinks, are undergoing the most severe declines over widespread areas. Very roughly, bobolinks still number about 11 million, but we are losing hundreds of thousands of bobolinks each year. Other Neotropical migrants are also experiencing sharp declines, including the painted bunting, cerulean warbler, and golden-winged warbler. Cerulean warblers still number more than half a million but their numbers have diminished by 70 percent over the past twenty-five years. More than 50 percent of golden-winged warblers have been lost since 1966 and only about 210,000 remain.[19] Although these species are not in immediate danger of extinction, if the precipitous declines continue, extinction is inevitable.

The birds in most acute danger are the ones with restricted summer and winter ranges, with small populations and/or rapid population declines. Bachman's warbler, an inhabitant of Cuba in winter and bottomland southern swamps of the United States in summer, may already be extinct. Only about 2,100 Kirtland's warblers exist in the world, inhabiting the jack pine forests of Michigan in summer and the Bahama islands in winter. Between 1961 and 1971, the number of Kirtland's warblers dropped by 60 percent. One study showed that only two young fledged from twenty-nine nests and that 70 percent of the nests were parasitized by cowbirds. Conservationists took the drastic action of removing cowbirds from the area, reducing cowbird parasitism to 3 percent of nests. These and other measures have halted the steep decline of Kirtland's warbler, but the species still has not rebounded dramatically.[20]

THE GOOD NEWS is that many bird species have shown remarkable resilience in recent decades, their numbers steady or even increasing when taken as a whole across wide geographic areas. Some adaptable birds, such as American robins, northern mockingbirds, and American crows, can undergo population increases in areas where they breed in backyards and suburban neighborhoods. Tufted titmice have spread northward, aided by the availability of food and land-use changes as northeastern forests have regenerated on small farms abandoned in the early 1900s. Other woodland birds, such as pileated woodpeckers and scarlet tanagers, have also benefited from tracts of regenerating forest in the Northeast. However, these long-term changes have simultaneously put other birds at a disadvantage. For example, brown thrashers and golden-winged warblers — inhabitants of shrubby fields, forest clearings, and forests edges — have become less abundant as forests have matured.[21]

Conservationists are working not only to reverse declines

of the most threatened bird species but to help prevent other birds from becoming threatened in the first place. In a landmark publication, *North American Landbird Conservation Plan,* Partners in Flight evaluated the vulnerability of all North American landbirds to help identify conservation priorities even for those species not facing imminent extinction. This vision for bird conservation involves planned stewardship that will help make sure birds are protected from threats that could be around the corner. The authors emphasize that the passenger pigeon, once the most numerous bird, was driven to extinction in the span of a human lifetime. Passenger pigeons once numbered about five billion, according to some estimates. In 1831, John James Audubon described a mile-wide flock that took three days to pass overhead. In 1878, passenger pigeons were still so abundant that hunters in Michigan easily slaughtered three hundred tons for the meat market. A mere thirty-six years later, the species was gone, a victim of merciless hunting and destruction of the beech forests where it nested.

Although there are no easy solutions for reversing declines of migratory songbirds, there are numerous everyday ways to help make their journey safer. As consumers, homeowners, and birders, we make decisions that can save individual birds' lives, and, collectively, may make a difference for their populations.

The choices that we make as consumers reverberate all the way to the boreal forests of Canada and the tropical forests of Latin America. Some 1.5 billion acres of boreal forest blanket an area from Alaska to Newfoundland, about one fourth of the earth's intact forests. Some three hundred bird species breed there, including about one third of all migrants that pass through the southern United States in spring.[22] Several warblers named after locales in the United States breed in the Canadian boreal forest, including most Tennessee warblers, Connecticut warblers, Philadelphia vireos, and Cape May warblers.

Many trees from these habitats are logged for such consumer products as paper towels, tissues, and toilet paper. Kimberly-Clark, the maker of Kleenex, Scott, Cottonelle, and Viva products, uses 1.1 million cubic meters of trees from Canada's boreal forest every year, according to the Natural Resources Defense Council. Other companies also use pulp from the boreal forests, logging all together some half a million acres of boreal forest every year. The Natural Resources Defense Council and other conservation groups are pressuring tissue products companies to reduce their consumption of virgin wood and encouraging consumers to buy recycled products. (Kleenex and Scott contain no recycled wood pulp, in contrast with the Canadian company, Cascades, which uses 96 percent recycled fiber.) Consumers can also help stop wasteful clear-cutting by taking their names off of mailing lists for catalogs. Each year about eight million tons of trees are turned into catalogs, according to Audubon. More than one billion catalogs, largely from the boreal forest, are produced each year for Land's End, J. Crew, Victoria's Secret, J.C. Penney, and L.L. Bean, according the National Wildlife Federation.

Coffee drinkers can help make a difference for migratory songbirds on their wintering grounds. Traditionally, coffee farmers grow their crops beneath the shady canopy of forest trees. In Mexico, more than 150 bird species inhabit shade-grown coffee and cacao plantations. Throughout

Central America, the Caribbean islands, and Colombia—regions important for migratory birds—there are more than 6.7 million acres of shade coffee plantations that provide relatively good wintering habitats. However, many farmers are now growing new coffee varieties that yield greater harvests under full sun. In Colombia and Mexico, sun coffee plantations support 94 to 97 percent fewer bird species than shade-grown coffee farms.[23] Although shade-grown coffee is often more expensive because it is more difficult to produce, buying a cup of "bird-friendly" coffee pays off in at least two ways—by making it possible for farmers to grow their crops beneath trees inhabited by birds, and by making a better-tasting brew, the result of slowly ripened beans.

Whether purchasing a light bulb or a car, we can help songbirds by purchasing items that do not contribute as much to acid rain and global warming. If everyone in the United States changed their household light bulbs for compact fluorescent bulbs, annual carbon dioxide emissions would decrease by about 125 billion pounds—and consumers would gain twenty-five dollars in energy savings over the life of each bulb. Purchasing energy-efficient vehicles also helps reduce air pollution and create demand for the industry to improve its standards. If all sport utility vehicles, pickups, and minivans improved their fuel economy by just seven miles per gallon, carbon dioxide emissions would diminish by at least four hundred thousand tons *every day*.[24] By improving energy efficiency in our homes and selecting energy-star appliances, we also save money in the long-term and buy time against global warming.

A few improvements around our homes and workplaces

can also have immediate payback for reducing hazards to migratory birds. Place bird feeders away from windows to prevent birds from flying into them and break up window reflections using decorative decals. Several major cities now have programs in place to reduce bird fatalities from collisions with buildings. In New York, the Project Safe Flight campaign asks workers to turn off external lights after midnight and to shield the glow from interior lights. In Chicago, the local Audubon chapter initiated a Lights Out program to encourage people to turn off decorative lights on tall buildings during spring and fall migration. This gesture is saving the lives of tens of thousands of migratory birds every year, according to the Chicago Field Museum.

Improving habitats in your own neighborhood can also give a boost to migratory birds. When Princeton University biologist Martin Wikelski and colleagues radio-tracked Swainson's thrushes and hermit thrushes during spring migration, three of the six birds landed within fifty-five yards of houses. The birds remained there all day, including one bird that found the food it needed in a few trees near where it had landed in Chicago. Planting trees and shrubs in yards and parks will provide places for the insects that migratory birds depend on as they migrate north. In autumn, songbirds eat fruit from trees and shrubs to build up their fat reserves for their long migration. Shrubs that yield berries in fall can provide an important source of fuel for migrating songbirds.[25] Even tiny pieces of nature are important for birds crossing fragmented landscapes.[26]

To help make neighborhood habitats safer for birds, the American Bird Conservancy recommends keeping cats indoors. Spread the word to help raise awareness among friends and neighbors that keeping cats inside can save the lives of birds and reduce deaths and injuries to cats from vehicles, coyotes, and other hazards.

Those who live near fields where songbirds nest can ask landowners to consider delaying mowing in spring and summer for as long as possible, preferably until August, to give the nestlings time to fledge. The U.S. Department of Agriculture has provided a boon to grassland birds and other wildlife through the Conservation Reserve Enhancement Program, which pays farmers to stop mowing and plowing altogether. More than 33 million acres of grasslands have been set aside on environmentally sensitive land for terms of ten or fifteen years.

Wild birds need advocates at all levels, and there are many ways to help, whether by becoming involved in local efforts to preserve land or supporting international work to save, restore, and manage natural habitats. Each year, 46 million birders in the United States spend $32 billion per year on wildlife watching. In 2001, these expenditures generated $85 billion in economic benefits for the nation, according to the U.S. Fish and Wildlife Service.[27] Birders are already playing an important economic role through recreational wildlife watching and belong to a key constituency that can help support conservation.

Numerous nonprofit organizations are conducting groundbreaking work on behalf of bird conservation and rely on financial help from members and donors. For example, The Nature Conservancy acquires land for preservation and is helping to protect about 15 million acres in the United States. The conservancy's Migratory Bird Program is focusing on stopover habitats for migratory birds along the Gulf Coast and a network of sites for migratory and resident grassland bird species in the western Great Plains. Through the "Parks in Peril" program, The Nature Conservancy has also helped bring conservation infrastructure and management to thirty-seven parks in Latin America encompassing more than 28 million acres of habitat.

Campaigns led by the American Bird Conservancy are advocating for better protections for birds against pesticides, cats, collisions with towers and buildings, and other threats. The Cornell Laboratory of Ornithology is documenting the impact of the caged-bird trade on Neotropical migratory songbirds, pushing for more stringent regulations, and working with local organizations to help raise awareness in local communities about the value of wild birds.

As birders, we can also contribute to knowledge of migratory birds through citizen-science projects designed to gather information from wide geographic areas, providing data that scientists need to understand bird migration and changes in population numbers through time. These projects may be as simple as counting birds at your feeder and reporting the results or as involved as conducting intensive surveys for breeding birds. Becoming part of the effort is as straightforward as consulting appendix B and contacting a project that suits your level of interest.

Finally, one thing that all birders can do is to share their appreciation of birds with a child, neighbor, friend, or relative. Show someone a woodland filled with springtime migrants, bend down for a moment to look at a nest in the grass, share a story or two about the phenomenal travels of songbirds throughout the year. Anyone who learns these secrets cannot help but value birds and place importance on protecting them.

What would it mean if we lost the bobolink? Thousands of years of evolution created an exquisite creature that sheds its plumage with the seasons. It is bright yellow and bold black in spring, streaked with earthy brown in autumn. The male's ebullient songs have embellished summer days for as long as bobolinks have had grasslands to sing in. Wrote Thoreau in *Walden*, "It is as if he touched his harp within a vase of liquid melody and when he lifted it out the notes fell like bubbles from the trembling string."[28] Within a bobolink is the innate knowledge and physiological stamina to roam the continents, but the bobolink returns faithfully each year to nest in a place that may or may not be there, depending on human whim. To lose this bird, or any other, is an irreversible tragedy—one that we can avoid if we take steps to ensure there are places for birds and other wildlife.

When migratory birds cross our paths, they compel us to stop and marvel at their beauty, their fluid lives in four seasons and distant places, and the ecological intricacies they require to survive. As the miracle of migration continues, the arrival of songbirds each spring is a cause for celebration, for summer would not be the same without them. In autumn, though, we cannot keep them; the songbirds grow restless and depart, leaving us with emptier winter days and a quiet reminder that we should not take them for granted.

$\mathcal{A}$CKNOWLEDGMENTS

As I wrote this book, three spring migrations and two fall migrations came and went. I am grateful to the many people who joined me throughout the journey. I thank my editor at Walker & Company, Jackie Johnson, for developing this book about songbird migration and asking me to write it. I have appreciated her guidance and support every step of the way. This book was also made possible in part by support from the Cornell Laboratory of Ornithology and by the encouragement of Allison Wells, former director of Communications and Marketing, who allowed me the freedom to conduct research for the book as part of my work there.

I give my heartfelt thanks to the many scientists and bird experts who took time to share their stories with me. I especially thank Sidney Gauthreaux, John Arvin, Bob Russell, Zoltán Németh, Bill Cochran, Martin Wikelski, Kurt Fristrup, Andrew Farnsworth, Dick Holmes, Nick Rodenhouse, Bob Beason, Kent McFarland, Chris Rimmer,

Michael O'Brien, Bill Evans, and Pete Marra, who talked with me at length about their work. Dick Holmes, Nick Rodenhouse, Michael Bradstreet, Michael O'Brien, Zoltán Németh, and the crew at Johnsons Bayou also hosted my visits to their field sites. For the sections on where to go bird watching, I greatly appreciated the perspectives of Bob Sargent, Starr Saphir, Paul Jorgensen, Sarah Rupert, Bob Barnes, Sheri Williamson, Rich Stallcup, Pete Dunne, Chris Wood, and David Bacab.

I thank Frank Moore, Russell Greenberg, Charles Walcott, and Jon McCracken for reviewing certain chapters of the manuscript, correcting errors when I went astray, and making suggestions for improvement. I have done my best to present accurate information throughout, but any remaining mistakes are entirely mine.

I am fortunate to have so many knowledgeable colleagues at the Cornell Laboratory of Ornithology who read portions of the manuscript and whose ideas, advice, and criticisms improved this book. For this, I thank David Bonter, Brian Sullivan, Wesley Hochachka, Stefan Hames, Kevin Mc-Gowan, Irby Lovette, Chris Tessaglia-Hymes, Jim Lowe, and Ken Rosenberg. Diane Tessaglia-Hymes, Jennifer Smith, and Patricia Leonard provided perspectives or assistance at various stages of the book. Many others at the Lab inspired the sections on bird watching, citizen science, and conservation through their own work in scientific research, education, and outreach.

I give special thanks to my parents, Berbie and David Chu, who have always nurtured my passion for birds and for writing. During this project, they generously spent vacations helping to care for their two grandchildren while

I worked on the manuscript or checked out birding hot spots.

I owe my greatest thanks to my birding partner and husband, Mark Chao. He gave freely of his own time, reading drafts, taking care of our two young children while I worked nights, weekends, and vacations, and postponed many birding outings we might have spent together. I have benefited from his help and insights in ways too numerous to count.

Appendix A
Songbird Migration Hot Spots

Migratory Songbird Spectacles

Gulf Coast

Alabama Coastal Birding Trail, Alabama

This birding trail winds through two counties and includes places known for migratory songbird fallouts, such as Dauphin Island and Fort Morgan. The Alabama Coastal BirdFest in October celebrates migration with birding trips, seminars, and workshops.

Alabama Gulf Coast Convention and Visitors Bureau: (800) 745-7263

http://www.alabamacoastalbirdingtrail.com

Gulf Islands National Seashore, Florida and Mississippi

Stretching across some 160 miles, this national seashore is an important stopover area for migratory birds. It includes

Fort Pickens and the Naval Live Oaks Reservation, hot spots during spring and fall migration.

Park visitor information, Florida: (850) 934-2600
http://www.nps.gov/guis

DRY TORTUGAS NATIONAL PARK, FLORIDA
Seven islands of coral reef and sand are a welcoming rest-stop for exhausted birds that have crossed the Gulf of Mexico in spring and for southbound migrants that are waiting for favorable weather before flying over the Gulf in fall.

Dry Tortugas National Park: (305) 242-7700
http://www.nps.gov/drto

GRAND ISLE BIRDING TRAIL, LOUISIANA
Grand Isle is a barrier island with oak-hackberry forests that sometimes overflow with songbirds during migration. The Migratory Bird Celebration in April includes bird watching tours and presentations.

Grand Isle Birding Trail: http://grandisle.btnep.org

PEVETO WOODS SANCTUARY, LOUISIANA
This chenier is a forty-acre hot spot for songbird migration. Researcher Frank Moore conducted mist-netting studies there during the 1980s.

Baton Rouge Audubon Society Sanctuaries:
http://www.braudubon.org/sanctuaries.asp

HIGH ISLAND BIRD SANCTUARIES, TEXAS

High Island is a premier location for songbird fall-outs during migration. Houston Audubon manages four sanctuaries in an "island" of woodlands surrounded by marshes.

Houston Audubon: (713) 932-1639

http://www.houstonaudubon.org

SOUTH PADRE ISLAND BIRDING AND
NATURE CENTER, TEXAS

Dune meadows, salt marshes, and thickets of shrubs and trees attract more than three hundred bird species to this island. The center has nature trails that are free and open to the public.

South Padre Island Convention and Visitor's Bureau: (956) 761-3005

http://www.worldbirdingcenter.org

Lakeshores

LESSER SLAVE LAKE BIRD
OBSERVATORY, ALBERTA

During migration, birds funnel through Lesser Slave Lake Provincial Park as they avoid crossing water and the nearby Marten Mountain. The observatory offers bird banding demonstrations for visitors. The Songbird Festival in June includes guided bird walks through the boreal forest to look for Neotropical migrants.

Lesser Slave Lake Bird Observatory:
http://www.lslbo.org

WHITEFISH POINT BIRD
OBSERVATORY, MICHIGAN

Located on a spit that extends into Lake Superior, the Whitefish Point Observatory is adjacent to the Whitefish Point Management Unit of the Seney National Wildlife Refuge. During peak migration weekends, the observatory offers bird banding demonstrations, educational programs, and bird walks.

Whitefish Point Bird Observatory:
http://www.wpbo.org

BRADDOCK BAY BIRD OBSERVATORY, NEW YORK

The observatory operates the Kaiser-Manitou Beach Banding Station on the south shore of Lake Ontario in spring and fall. The staff encourages visitors to come watch birds being banded and to learn about bird migration and research. Call in advance to check mist-netting schedules.

Braddock Bay Bird Observatory: (585) 234-3525
http://www.bbbo.org

CRANE CREEK STATE PARK AND
MAGEE MARSH, OHIO

Songbirds stop over in this park on the southern shore of Lake Erie. The boardwalk in the adjoining Magee Marsh is especially renowned for warbler watching.

Ohio Department of Natural Resources:
http://www.dnr.state.oh.us/parks/parks/cranecrk.htm

LONG POINT BIRD OBSERVATORY,
ONTARIO (SEE PAGES 167–69)

POINT PELEE NATIONAL PARK, ONTARIO (SEE
PAGES 79–81)

Desert Oases

SAN PEDRO RIPARIAN NATIONAL CONSERVATION
AREA AND HUACHUCA MOUNTAINS, ARIZONA (SEE
PAGES 130–32)

BUTTERBREDT SPRING, CALIFORNIA (SEE PAGE 127)
http://www.recreation.gov/detail.cfm?ID=618

JOSHUA TREE NATIONAL PARK, CALIFORNIA
Within the park, five oases provide refuge for migrants
passing through the Colorado and Mojave deserts.

Joshua Tree National Park: (760) 367-5500

http://www.nps.gov/jotr/activities/birding/birding.html

YAQUI WELL, ANZA-BORREGO DESERT STATE
PARK, CALIFORNIA (SEE PAGES 77–79)

MALHEUR NATIONAL WILDLIFE
REFUGE, OREGON
Migratory birds are attracted to the refuge, an oasis of
wetlands in the high desert of southeastern Oregon. The

trees may be brimming with western tanagers, Bullock's orioles, and lazuli buntings during spring fallouts.

Malheur National Wildlife Refuge: (541) 493-2612
http://www.fws.gov/pacific/malheur

Urban Oases

THE MAGIC HEDGE, COOK COUNTY, ILLINOIS
Located along the shore of Lake Michigan on the north side of Chicago, the Magic Hedge is a small oasis of trees, shrubs, and grasses surrounded by urban development, farmland, and lake water.

Chicago Wilderness Magazine: http://chicagowilderness
mag.org/issues/spring1998/IWmagichedge.html

"THE MIGRANT TRAP," HAMMOND, INDIANA
Migrants concentrate in this six-hundred-meter wooded area in an urban area near the shore of Lake Michigan.

Hammond Marina: (219) 659-7678
http://www.indianaaudubon.org/guide/sites/migtrap.htm

MOUNT AUBURN CEMETERY, CAMBRIDGE, MASSACHUSETTS
Surrounded by urban development, the Mount Auburn Cemetery is an oasis of greenery with more than ten miles of roads and paths for visitors to travel. In spring, when numerous fall migrants come through, local birding clubs maintain a sightings board at the entrance.

http://www.nationalgeographic.com/destinations/
Boston/Boston_Area_Bird_watching.html

CENTRAL PARK, NEW YORK CITY (SEE PAGES 82–84)

Coasts and Corridors

BIG SUR ORNITHOLOGY LAB, CALIFORNIA

Located in Andrew Molera State Park on the Pacific Coast, the Big Sur Ornithology Lab bands thousands of migratory songbirds. The staff offer banding demonstrations on some days of the week.

Big Sur Ornithology Lab: (831) 624-1202

http://www.ventanaws.org/lab.htm

PRBO's PALOMARIN FIELD STATION, CALIFORNIA (SEE PAGES 169–71)

CAPE MAY, NEW JERSEY (SEE PAGES 165–67)

KIPTOPEKE STATE PARK, VIRGINIA

The Coastal Virginia Wildlife Observatory bands songbirds and raptors during migration at the tip of the Delmarva peninsula. The Eastern Shore Birding Festival in October features workshops, guest speakers, educational activities, and boat tours.

Kiptopeke State Park: (757) 331-2267

Coastal Virginia Wildlife Observatory:
http://www.dcr.state.va.us/parks/kiptopek.htm

Islands

Monhegan Island, Maine

Twelve miles off the coast of Maine, the rocky coastline, low-growing island vegetation, and spruce forests of Monhegan Island attract songbirds during migration.

http://gorp.away.com/gorp/location/me/monhegan4.htm

Grand Manan Island, New Brunswick

From its rocky shores to the cliffs, bogs, and coniferous forests, this island in the Bay of Fundy attracts a diversity of birds. Warblers, sparrows, and other songbirds take refuge on the island in spring and fall. The surrounding waters are home to shearwaters, razorbills, Atlantic puffins, and other pelagic birds.

Grand Manan Island Visitor Information Center: (506) 662-3442

http://www.tourismnewbrunswick.ca

Block Island, Rhode Island

Block Island is twelve miles from the Rhode Island coast and is along a major migratory flyway. It is known for high concentrations of migratory songbirds in fall. Many of the birds are young and inexperienced, apparently deciding to stop in the coastal shrubs and trees after finding themselves too far over water on their route south. The Block Island Banding Station (located along the Clay Head Trail) welcomes visitors during spring and fall migration. Since it is a private residence, please contact Kim Gaffett at kimg@riconnect.com to inquire about access and schedules. The Audubon Society of Rhode Island leads field trips and programs during Block Island Birding Weekend in fall. A fee is required to cover lodging, meals, travel, and staffing.

Audubon Society of Rhode Island: (401) 949-5454

http://www.asri.org

Block Island National Wildlife Refuge:
http://www.fws.gov/refuges

SUMMER BIRDING: DIVERSE HABITATS, DIVERSE BIRDS

SOUTHERN SIERRA NEVADA AND KERN RIVER VALLEY, CALIFORNIA (SEE PAGES 127–29)

SOUTHEASTERN ARIZONA (SEE ALSO SAN PEDRO RIPARIAN NATIONAL CONSERVATION AREA AND HUACHUCA MOUNTAINS ON PAGES 130–32)

In a day's drive, visitors to southeastern Arizona can see diverse bird species inhabiting deserts, riparian areas, and mountains. The Southeastern Arizona Bird Observatory has information on birding hot spots, guided walks, and spring and fall bird festivals.

Southeastern Arizona Bird Observatory:
http://www.sabo.org

Pawnee National Grassland and Rocky Mountain National Park, Colorado

Only about fifty miles separates portions of the Pawnee National Grassland from the Rocky Mountain National Park. Birders traveling from the grasslands to the mountains see a shift in bird communities from lark sparrows, lark buntings, horned larks, and other grassland birds to montane species such as mountain bluebirds, Clark's nutcrackers, green-tailed towhees, and Cassin's finches. The Rocky Mountain Bird Observatory, about sixty miles southeast of Rocky Mountain National Park, offers educational programs, including programs about migration and bird banding as well as nests and niches.

Pawnee National Grassland: (970) 346-5000. Request a free brochure, "Birding on the Pawnee."

Rocky Mountain National Park: (970) 586-1206

http://www.nps.gov/romo

Rocky Mountain Bird Observatory educational programs: (303) 637-9220

http://www.rmbo.org

Pine to Prairie Birding Trail, Minnesota

Visitors to northern Minnesota can watch birds in grasslands, tallgrass prairies, mountains, and boreal, aspen, northern hardwood, and coniferous forests. The annual Festival of Birds takes pace in the Detroit Lakes area in May.

http://www.mnbirdtrail.com

Niobrara Birding Trail, Nebraska

This birding trail includes areas in between eastern deciduous forests and the coniferous forests of the Rocky Mountains. The region has a mix of eastern birds (wood thrush, black-and-white warbler, scarlet tanager) and western species (Bullock's oriole and lazuli bunting). The varied habitats include grassland birds such as lark buntings and McCown's and chestnut-collared longspurs. Inhabitants of coniferous forests include pinyon jays and plumbeous vireos.

Nebraska Birding Trails Project:
http://www.nebraskabirdingtrails.com

Riding Mountain National Park, Manitoba

More than 250 species of birds have been found in Riding Mountain National Park and 160 species nest there regularly. The diverse habitats include eastern deciduous forest, northern boreal forest, and grasslands.

Riding Mountain National Park: (204) 848-7275
http://www.pc.gc.ca/pn-np/mb/riding/index_e.asp

THE TROPICS

Belize

Chan Chich Lodge

Some 350 bird species have been recorded on the private reserve surrounding the lodge. More than nine miles of trails wind through a rainforest alive with birds and other wildlife, including jaguars, black howler monkeys, and spider monkeys.

Chan Chich Lodge U.S. office: (800) 343-8009

http://www.chanchich.com

LA MILPA FIELD STATION

The field station, located near a Mayan archaeological site, is surrounded by nine trails. Common bird sightings include toucans, crested guans, hummingbirds, and euphonias. The station is run by the Programme for Belize, a Belizean nonprofit conservation organization.

Programme for Belize: 011-501-227-5616

http://www.pfbelize.org/lamilpa.html

Costa Rica

LA SELVA BIOLOGICAL STATION

Managed by the Organization for Tropical Studies, La Selva is a 1,600-acre reserve in lowland tropical forests and a core conservation unit of the Cordillera Volcanica Central Biosphere Reserve. More than four hundred bird species have been found here. The biological station welcomes visitors to stay at the lodge if space is available. The station offers guided walks, birding tours, boat tours, and workshops, including Birdwatching 101. Wintering Neotropical migrants include Tennessee warblers, chestnut-sided warblers, and Baltimore orioles (abundant), and red-eyed vireos, wood thrushes, yellow warblers, northern waterthrushes, Kentucky warblers, and mourning warblers (common).

La Selva Biological Station: 011-506-766-6565
http://www.ots.duke.edu/en/laselva

SELVA BANANITO LODGE

More than three hundred species have been found in the two thousand-acre private reserve and the immediate vicinity. Habitats include forests, banana fields, and cacao plantations. The family-owned business offers bird tours, horseback riding, tree-climbing lessons, and an opportunity to observe birds from the canopy. Located along a major migration route about nine miles from the Pacific coast, the reserve is used by sixteen passage and wintering Neotropical migrant wood-warbler species.

Selva Bananito Lodge: http://www.selvabananito.com

Jamaica

FROM MANGROVES TO MOUNTAINS

Some 265 species of birds have been recorded in Jamaica, with 27 endemic species. Many tour groups arrange to stay at Marshall's Pen, a private ranch near Mandeville, fifty-five miles west of Kingston. From there, visitors often travel to the mangroves, lagoons, and marshes of the Black River Morass in southwest Jamaica, the rugged forests of Cockpit Country, the acacia scrub of Portland Ridge, and the rainforest of the Blue Mountains.

Mexico

COLIMA AND JALISCO

From Pacific beaches to forested volcanoes, this region has diverse habitats, with some four hundred species recorded. The ascent from lowlands to the slopes of the volcanoes is marked by changes in bird species, both resident and migratory.

Panama

CANOPY LODGE, PANAMA

In the crater of an extinct volcano, the forests and gardens of El Valle de Antón are home to spectacular birds, including forty-one species of tyrant flycatchers, twenty-four species of hummingbirds, four species of trogons, four species of motmots, and Neotropical migrants including Swainson's thrush, wood thrush, and fifteen warbler species.

Canopy Lodge: 011-507-264-5720

http://www.canopylodge.com

CANOPY TOWER ECOLODGE AND NATURE OBSERVATORY, PANAMA (SEE SOBERANÍA NATIONAL PARK ON PAGES 217–18)

SMITHSONIAN TROPICAL RESEARCH INSTITUTE, PANAMA

The research station on Barro Colorado Island is open to the public three days a week for guided tours. The three-hour walk ends at exhibits in the visitors' center, followed by lunch.

Smithsonian Tropical Research Institute: 011-507-212-8026

http://www.stri.org

Appendix B
CITIZEN-SCIENCE PROJECTS

IF YOU ENJOY watching birds, consider recording your counts or observations for a citizen-science project. Projects are fun and help scientists to study migration patterns, bird distributions, and changes in bird numbers through time. Some projects can take as little time as fifteen minutes per year if you choose (Great Backyard Bird Count). Some projects request that you count birds at your feeder once a week or once every other week during part of the year (Project FeederWatch); others leave the decision entirely up to you (eBird). The success of these projects relies on participation from many people. Each contribution adds more information for scientists to analyze, allowing them to make stronger conclusions about the biology and conservation of birds.

The list of projects below is a starting point, but check with your local birding club, Audubon chapter, bird observatory, or nature center for additional citizen-science projects that may need help in your region.

SPRING/SUMMER

The Birdhouse Network

When: Spring and summer

What's involved: Participants put up one or more nest boxes and record whether they remain empty or whether a cavity-nesting bird settles in. Throughout the breeding season, participants get a close-up look at life in the nest as they observe birds and check the nest for eggs and young.

What has been learned? Scientists have used the results to study geographical variation in clutch size, potential effects of global warming on egg-laying dates, and effects of pesticides on nesting success. The eggs of eastern bluebirds in the southern United States fail to hatch more frequently than in areas farther north, especially later during the nesting season in association with warmer temperatures.

Why contribute? Nest-monitoring observations help improve understanding of birds' breeding behaviors and the factors that influence nesting success.

How to get started: Visit http://www.birds.cornell.edu/ birdhouse for information about making or purchasing nest boxes and attracting cavity-nesting birds. The Web site also offers excellent summaries of the biology of cavity-nesting birds and reports on numerous results from past studies. Complete instructions and data entry are available through the Web site. A small fee is charged to help support the program (fifteen dollars, or twelve dollars for members of the Cornell Laboratory of Ornithology). Call (607) 254-2416 for more information.

Sponsoring organization: Cornell Laboratory of Ornithology

Birds in Forested Landscapes

When: Spring and summer

What's involved: Participants survey forested habitats for birds in need of conservation. They record the presence or absence of birds, along with information about location, habitat, breeding status, predators, and cowbirds.

What has been learned? Participants' data have helped show a link between acid rain and declining wood thrush populations. The data have also been used to provide land managers with guidelines for the conservation of thrushes and tanagers.

Why contribute? The records help researchers formulate conservation guidelines and understand how birds are affected by human impacts on the environment, such as forest fragmentation, acid rain, and mercury contamination.

How to get started: Visit http://www.birds.cornell.edu/bfl for background information and instructions. Participation is free. Call (607) 254-2413 for more information.

Sponsoring organization: Cornell Laboratory of Ornithology

Breeding Bird Atlas

When: Spring and summer

What's involved: Various organizations may coordinate Breeding Bird Atlases differently, but in general participants census birds in an atlas block of a few square miles several times during the breeding season. The blocks extend across the entire state, province, or territory, and are usually surveyed for a period of five years at a time.

What has been learned? The results are compiled and published in books or online publications. Scientists have used these records to examine how forest cover, habitat type, and climate influence the distribution and abundance of birds. For example, information from the Pennsylvania Breeding Bird Atlas helped document declines of bird species in Pennsylvania grasslands and wetlands.

Why contribute? Breeding bird atlases document information that can be used to understand the distribution of nesting birds in the landscape and to aid in conservation management plans.

How to get started: For links to Breeding Bird Atlases, visit http://www.bsc-eoc.org/links/links.jsp?page=g_atlas.

Sponsoring organization: North American Ornithological Atlas Committee

Breeding Bird Survey

When: Spring and summer

What's involved? Established in 1966, this comprehensive survey collects data on breeding birds from predetermined routes across the United States and Canada. Beginning about half an hour after sunrise, volunteers survey a twenty-five-mile route along roads or waterways. Every half-mile, volunteers stop and record all birds seen or heard in a three-minute period. In all, volunteers cover some fifty thousand miles in their surveys.

What has been learned? Data from the Breeding Bird Survey are widely used. Partners in Flight formulated the North American Landbird Conservation Plan using Breeding Bird Survey data for most of the species. By identifying the species undergoing the biggest declines, the plan outlined priorities for one hundred species and recommended conservation objectives based on the severity of decline and other factors.

Why contribute? The Breeding Bird Survey has helped scientists document long-term trends in breeding bird populations from coast to coast.

How to get started: For more information, visit http://www.pwrc.usgs.gov/bbs.

Sponsoring organizations: U.S. Geological Survey and Canadian Wildlife Service

Other Spring/Summer Citizen-Science Projects

Great Lakes Marsh Monitoring Program
(888) 448-2473; http://www.bsc-eoc.org/mmptell.html
Migration Monitoring (spring and fall)
http://www.pwrc.usgs.gov/birds/othbird.html#mig
National Marsh Bird Monitoring Program
(520) 626-8535
Neighborhood Nestwatch
(443) 482-2344; http://sio.si.edu/Nestwatch
North American Migration Count
http://community.gorge.net/birding/namcstasz.htm
Ontario Forest Bird Monitoring Program

(519) 826-2094; http://www.pwrc.usgs.gov/birds/
othbird.html#mig

Vermont Forest Bird Monitoring Program

(802) 457-2779, ext. 123; http://www.vinsweb
.org/cbd/FBMP.html

Nest Record Card Programs

Canadian Wildlife Federation Nest Record Scheme
(British Colombia, the Prairies, Ontario, Quebec, and
the Maritimes)

http://www.cwf-fcf.org/pages/wildresources/surveys/
survey34.htm

Louisiana Nest Record Program

http://appl003.lsu.edu/natsci/NestingBirds.nsf/
MainPage?OpenForm

Northwest Nest Record Program (Washington,
Oregon, Idaho, Montana, British Columbia)

(206) 543-1668; http://www.wos.org/NestRpts.htm

Ontario Nest Record Scheme

(416) 586-5523;
http://www.birdsontario.org/onrs/onrsmain.html

Prairie Nest Records Scheme (Alberta)

(780) 427-8124; http://fanweb.ca (click on "projects
and programs")

Project NestWatch (Canada)

(888) 448-BIRD; http://www.bsc-eoc.org/national/
nestwatch.html

Vermont Nest-Record Card Program

(802) 457-2779; http://www.vinsweb.org/cbd/vt_
nestrecord.html

AUTUMN/WINTER

Christmas Bird Count

When: One day, between December 14 and January 5.

What's involved: The Christmas Bird Count begins at midnight on a designated day and ends the following midnight. Participants work in teams to count as many bird species and individuals as possible in assigned sections of a count circle fifteen miles in diameter. Groups gather to compile their counts, and the tallies are combined with counts from across the United States, Canada, Latin America, and the Caribbean.

What has been learned? The Christmas Bird Count was founded in 1900 by ornithologist Frank Chapman as an alternative to the Christmas "Side Hunt," a competition to see which team could shoot the most birds and other animals. With more than one hundred years of data, the Christmas Bird Count has become a valuable source of information about changes in bird populations over time. Data from the Christmas Bird Count have helped track irruptions of boreal seed-eating birds. More recently, the counts have been used to document declines of American crows that appear to be related to West Nile virus. The long-term records have also shown how the ranges of birds have expanded or contracted through time, such as the northward spread of tufted titmice since 1902. The counts also track changes in numbers, such as declines of Bewick's wrens in the eastern United States since the 1930s.

Why contribute? As the longest-running data set on bird populations, the Christmas Bird Count helps scientists keep track of long-term population changes.

How to get started: The Christmas Bird Count is organized through local coordinators. Call (215) 355-9588 or visit http://www.audubon.org/bird/cbc for more information. A five-dollar donation is requested from participants.

Sponsoring organization: Audubon

Great Backyard Bird Count

When: Friday through Monday of President's Day weekend in February.

What's involved: The Great Backyard Bird Count is fun and easy for beginning and expert bird watchers alike. Participants count birds from locations of their choosing for at least fifteen minutes on one or more of the count days. After tallying the highest number of each bird species seen, they submit the results online at the Great Backyard Bird Count Web site, http://www.birdsource.org/gbbc. As the data pour in, participants can explore how their sightings fit in with what others are reporting in their own towns and around the United States and Canada. Summaries and maps show bird counts regionally or continent-wide. Participants can also contribute to a photo gallery showing close-up images of birds, both common and rare, that were photographed during the count.

What has been learned? Data from the Great Backyard Bird Count have helped reveal the dynamic migratory patterns of sandhill cranes, the rapid spread of the introduced

Eurasian collared-dove, and the movements of winter finches. Preliminary analyses show that robins avoid areas with snow cover.

Why contribute? During four days, reports from tens of thousands of participants provide a continentwide snapshot of winter bird populations. Within minutes of submitting results, participants can view maps and tables showing where the birds are being seen and can compare the counts with previous years.

How to get started: Go to http://www.birdsource.org/gbbc for complete instructions, data entry, and access to the results. There is no fee or registration. For more information, call (607) 254-2473.

Sponsoring organizations: Cornell Laboratory of Ornithology, Audubon, and Wild Birds Unlimited

Project FeederWatch

When: November–April.

What's involved: FeederWatchers keep their feeders filled in winter and count the birds for as little or as long as they like on two consecutive days every two weeks (if submitting data on paper forms) or as often as every week (if submitting data online). Participants report the highest number of each species seen during the two count days.

What has been learned? FeederWatchers have documented interesting migratory patterns of numerous bird species that winter in the United States and Canada, including biennial irruptions of species such as common redpolls, movements of pinyon jays that may be related to fire and/or drought, and movements of juncos to lower altitudes in some years. Scientists have also used the data to document seed preferences of birds (for results, visit http://www.birds.cornell.edu/programs/AllAboutBirds/attracting/feeding/food_pref).

Why contribute? By sending in counts, FeederWatchers help scientists understand the distribution, movements, and population trends of feeder birds. Participants can contribute to and access online photo galleries, rare bird reports, tallies, and an e-mail discussion group. They also receive a newsletter sharing highlights from each season, a Common Feeder Birds poster, and *BirdScope,* the Cornell Lab of Ornithology's newsletter, reporting results from Project FeederWatch and other citizen-science projects.

How to get started: For more information or to sign up, U.S. residents can visit http://www.birds.cornell.edu/pfw or call (800) 843-2473. Canadians can visit http://www.bsc-eoc.org/national/pfw.html or call (888) 448-2473. To offset costs of project materials and mailings, a fee of fifteen dollars (twelve dollars for Lab of Ornithology members) is requested from participants in the United States and thirty-five dollars from participants in Canada.

Sponsoring organizations: Cornell Laboratory of Ornithology and Bird Studies Canada

YEAR-ROUND

eBird

When: Any time.
What's involved: Participants keep a checklist of the birds they see—whether at feeders or while birding anywhere in

the United States, Canada, and Mexico. They enter sightings online at eBird, http://www.ebird.org. These observations become part of a central database accessible by anyone with Internet access, including scientists, educators, students, and bird watchers. Interactive tools make it easy to keep lists and explore sightings using graphs, maps, and charts.

Regional projects are available in some states and countries, including Vermont eBird, http://www.ebird.org/VINS; Mass Audubon eBird, http://www.ebird.org/MassAudubon; Texas eBird, http://www.ebird.org/tx; and a VerAves in Mexico, http://www.ebird.org/averAves.

What has been learned? Data from eBird have been used to track the timing and migratory routes of warblers, hummingbirds, and hawks and to document patterns of irruptive movements of winter finches.

Why contribute? If you enjoy watching birds, entering your records into eBird can have both personal and scientific rewards. Using eBird, you can keep track of your own bird sightings over the years and use eBird to learn how the composition and number of birds changes at your favorite locales over time. Your records can also help document seasonal changes such as the progression of migrants into and out of your area each year. The observations become part of a historical database that can be used by scientists to document migratory movements of birds. As a year-round citizen-science project, eBird is helping to document the timing and geographic distribution of migrants across the continent.

How to get started: Go to http://www.ebird.org for instructions, data entry, and results. The Web site also offers seasonal feature stories about birds, birding, and interesting trends revealed by eBird data. Participation is free.

Sponsoring organizations: Cornell Laboratory of Ornithology and Audubon

House Finch Disease Survey

When: Year-round.

What's involved? Participants record how many house finches they see (even if none) and whether the birds have symptoms of house finch eye disease.

What has been learned? Since 1994, participants have documented house finch eye disease from its origins in the Washington, D.C., area, and its spread to the West Coast. Combined with the Christmas Bird Count, these data indicate that an estimated 180 million fewer house finches existed in 2000 than if the disease had never occurred.

Why contribute? Scientists are studying the spread of disease and the dynamics of epidemics. The results could be valuable in understanding other diseases that affect wildlife and even humans.

How to get started: Visit http://www.birds.cornell.edu/hofi to learn how to recognize the symptoms of house finch eye disease. Complete instructions and data entry are available at the Web site. Participation is free. For more information, call (607) 254-2469.

Sponsoring organization: Cornell Laboratory of Ornithology

Urban Bird Studies

When: Year-round.

What's involved? Participants choose from five projects:

Birds in the City, PigeonWatch, Crows Count, Dove Detectives, or Gulls Galore. They observe and count birds and fill out data sheets to report their totals. The projects are ideal for people of all ages and levels of birding expertise.

What has been learned? Participants of PigeonWatch have collected information about the coloration of pigeons in different regions. They found that more pigeons in northern and western areas have dark plumage.

Why contribute? Scientists want to know more about where birds live in cities, which species seem to thrive in cities, and whether birds behave differently in cities than elsewhere.

How to get started: Visit http://www.urbanbirds.org for background information and instructions. Participation is free. For more information, call (607) 254-2455.

Sponsoring organization: Cornell Laboratory of Ornithology

Banding Stations

Many banding stations and bird observatories accept volunteer help. These Web sites provide lists of regional banding stations:

Banding Associations and Bird Observatories listed by USGS Bird Banding Laboratory

http://www.pwrc.usgs.gov/BBL/manual/birdobs.htm

Bird Observatories, USA and Canada, listed by BIRDNET

http://www.nmnh.si.edu/BIRDNET/OBSERVATORY
.html

Canadian Migration Monitoring Network

http://www.bsc-eoc.org/national/cmmn.html

Appendix C
RECOMMENDED RESOURCES

CHAPTER 4. WITNESSING THE SPRING SPECTACLE

Able, K. P. "Migration Biology for Birders." *Birding*, April 1991.

Clemson University Radar Ornithology Laboratory, http://virtual.clemson.edu/groups/birdrad. This Web site from Sidney Gauthreaux's laboratory includes information about radar, weather, and bird migration.

College of DuPage Web site, http://weather.cod.edu/analysis/analysis.radar.html.

National Center for Atmospheric Research Real-Time Weather Data, http://www.rap.ucar.edu/weather/model.

New Jersey Audubon Society. "Preserving Oases Along the Flyway." http://www.njaudubon.org/Education/Oases/RadWork.html. These pages show useful examples of radar images and how to interpret them.

The Weather Channel Web site, http://www.weather.com.

CHAPTER 7. HOW TO FIND AND MONITOR BIRD NESTS
All About Birds, http://www.allaboutbirds.org.
This free resource from the Cornell Laboratory of
Ornithology includes tips about providing feeders,
food, and backyard habitats for breeding birds.
The Online Bird Guide profiles North America's
most common bird species, including details about
nesting habits.

The Birdhouse Network, http://www.birds.cornell.
edu/birdhouse. This citizen-science project from the
Cornell Laboratory of Ornithology offers free profiles
of cavity-nesting species and information on
constructing and monitoring nest boxes.

The Birds of North America Online, http://bna.birds.
cornell.edu. A subscription to this electronic resource
gives you access to the most up-to-date and
authoritative accounts of North American bird species,
including detailed sections on breeding behavior and
biology.

Nest Box Cam, http://www.birds.cornell.edu/
birdhouse. This Web site from The Birdhouse Network
transmits lives images of nesting birds via the
Internet. You can peek in on the nests of bluebirds,
chickadees, wrens, swallows, or prothonotary warblers
during the breeding season. Archived images allow you
to take a look back at the entire sequence as birds
build nests, lay eggs, and raise their young.

CHAPTER 9. LOOKING AND LISTENING FOR AUTUMN MIGRATION

Dunn, Jon, and Kimball Garrett. *A Field Guide to Warblers of North America*. Boston, Mass.: Houghton Mifflin, 1997. This guide to warblers includes illustrations of all major plumages variations, as well as information on identification, timing of migration, and vocalizations.

Elliott, L., D. Stokes, and L. Stokes. *Stokes Field Guide to Bird Songs: Eastern Region*. New York: Time Warner Audio Books, 1997.

Cloven, K. J., D. Stokes, and L. Stokes, *Stokes Field Guide to Bird Songs: Western Region*. New York: Time Warner Audio Books, 1999.

Evans, W., and M. O'Brien. *Flight Calls of Migratory Birds* CD-ROM. This CD-ROM is a reference guide to flight calls of 211 migratory landbirds in eastern North America, including audio recordings, spectrograms, and information on migration and calling behavior. To purchase, visit http://www.oldbird.org.

Kaufman, Kenn. *Birds of North America*. Boston: Houghton Mifflin Co., 2000. This field guide to North American birds includes helpful photographs showing warblers and other songbirds in spring and fall plumages.

Old Bird, http://www.oldbird.org. This Web site from Bill Evans provides all the instructions you need to set up the equipment for monitoring night flight calls of migrating birds.

Raven Lite sound analysis software. Produced by the Cornell Laboratory of Ornithology, Raven Lite is designed for anyone who wants to learn more about bird sounds by using spectrograms and other sound analysis tools. To download the free software, visit http://www.birds.cornell.edu/brp.

Sibley, David Allen. *The Sibley Guide to Birds.* New York: Alfred A. Knopf, 2000. Sibley's guide to North American birds includes useful descriptions of vocalizations and illustrations that are helpful in distinguishing songbirds in fall plumage.

CHAPTER 11. BIRDS OF NORTHERN WINTERS

Goodwin, M. L. *Birding in Venezuela.* Caracas, Venezuela: Sociedad Conservationista, Audubon de Venezuela, 1997.

Howell, S. N. G. *A Bird-Finding Guide to Mexico.* Ithaca, N.Y.: Comstock, 1999.

Sekerak, A. D. *A Travel and Site Guide to Birds of Costa Rica, with Side Trips to Panama and Nicaragua.* Edmonton: Lone Pine, 1996.

Taylor, K. *A Birder's Guide to Costa Rica.* n.p.: Keith Taylor Birdfinding Guides, 1993.

Valqui, T. *Where to Watch Birds in Peru.* Lima, Peru: Thomas Valqui, 2004.

Victor Emmanuel Nature Tours (VENT): (800) 328-8368; http://www.ventbird.com.

Wheatley, N., and D. Brewer. *Where to Watch Birds in Central America and the Caribbean.* London: A&C Black, 2001.

White, A. W. *A Birder's Guide to the Bahama Islands*. Colorado Springs, Col.: American Birding Association, 1998.

WINGS: (888) 293-6443; http://www.wingsbirds.com.

For a list of additional tour companies, visit www.birding.com/TourCompanies.asp.

Appendix D
ABOUT THE CORNELL
LABORATORY OF
ORNITHOLOGY

THE CORNELL LABORATORY of Ornithology is a nonprofit, member-supported organization known for its innovative approach to bird study. Housed in the Johnson Center for Birds and Biodiversity in Ithaca, New York, the Lab is a hub for cutting-edge ornithological research, conservation, education, and outreach.

These longstanding traditions go back to 1915, when the Laboratory of Ornithology was founded by Arthur A. Allen, the first professor of ornithology in the United States. In addition to teaching courses at Cornell and pioneering the use of sound-motion picture film to study birds, Allen popularized bird watching through public lectures, articles, and radio shows. Olin Sewall Pettingill, director from 1960 to 1973, studied bobolinks and other birds during his long career and reached the public through bird films, articles, and books. Some ten thousand students graduated from his popular Home Study Course in Bird Biology. Charles Walcott, director from 1981 to 1995, conducted landmark experiments demonstrating that homing pigeons use magnetic fields to navigate. He reached millions of people through nationally televised shows about birds.

Since 1995, director John Fitzpatrick has ushered in an exciting new era that has extended the Lab's research, education, and conservation efforts around the globe:

- The Lab's scientists and engineers are developing technologies to detect and monitor elusive wildlife by recording sounds. They have launched autonomous recording units into the ocean to listen for whales, floated recording units on balloons to listen for endangered black-capped vireos and golden-cheeked warblers over inaccessible areas, and strapped the units onto trees to detect the sounds of one of the most rare and elusive birds in the world, the ivory-billed woodpecker. The same technology is being used to monitor nocturnally migrating songbirds at acoustic stations around the continent. The Lab also develops software for screening and analyzing sound recordings using spectrograms. Raven Lite is a special version of the software that birders and educators can use to explore the vocalizations of birds and other animals. In addition, the Lab's engineers are developing automated systems to help track wildlife, including migratory songbirds, with a new generation of radio transmitters.
- The Lab's Macaulay Library is a multimedia collection on animal behavior, including the world's largest archive of bird sounds and video.

From this collection, the Lab produces bird song audio guides and provides sounds and video images for bird-related resources, such as the All About Birds Web site, http://www.allaboutbirds. org; the *Birds of North America Online,* http://bna.birds.cornell.edu; and the Cornell Lab of Ornithology/Thayer Software CD-ROMs. Bird sounds and images are also available by request for a fee, which is necessary to offset processing costs.

- From basic research to applied conservation, Lab scientists and their graduate students are investigating a diverse array of topics, including nocturnal flight calls of songbirds, evolutionary history of warblers, vocal communication and functions of dawn song, dynamics of house finch eye disease, and DNA studies to help preserve the genetic diversity of bird populations. The conservation staff helped formulate the Partners in Flight Landbird Conservation Plan and serve on the species recovery teams for endangered birds. In Latin America and the Caribbean, they are documenting the impact of the caged-bird trade on migratory birds and are working with Cuban collaborators to census and monitor migratory birds and endemic species in Cuba.

- Educational and citizen-science programs engage people of all ages in learning about birds and participating in scientific study, from the Classroom BirdWatch curriculum to the updated version of Pettingill's Home Study Course in Bird

Biology. Tens of thousands of people participate in citizen-science projects, gathering observations of birds from across North America. Using these data, Lab scientists have identified seed preferences of feeder birds, documented breeding distributions of golden-winged warblers and cerulean warblers for conservation purposes, provided guidelines for conservation management of thrushes and tanagers, implicated acid rain in the decline of wood thrushes, and documented complex migratory patterns of birds.

- The Lab's information science team has made it possible to gather, retrieve, store, and disseminate information about birds in powerful new ways. The Great Backyard Bird Count and eBird make participant data available nearly instantaneously through interactive features such as maps and graphs. A special version of eBird has been launched in Mexico as aVerAves, the first online bird checklist monitoring program in Latin America. Meanwhile, the Lab has helped the Pennsylvania Breeding Bird Atlas go online, a step that has greatly speeded up the process of data entry and access to results. Ultimately, a new Avian Knowledge Network will allow easy access to millions of bird records currently stored by many kinds of projects at different ornithological institutions.

To improve the distribution of interpreted information, the information science team is developing *The Birds of North America Online* as a model for online scientific references. These "living publications" will be continuously updated through author input, community feedback, and automatic updates from new online sources.

- The Lab welcomes visitors to the Johnson Center for Birds and Biodiversity. The visitors' center has public exhibits, seminars, multimedia presentations, art exhibits, a bird-feeding garden, and more than four miles of trails and boardwalks in the Sapsucker Woods Sanctuary.

The Cornell Laboratory of Ornithology's programs are made possible through the support of friends and members. Members receive *Living Bird* magazine and the *BirdScope* newsletter, with the latest results from the Lab and its citizen-science projects. For more information or to become a member, visit http://www.birds.cornell.edu or call (800) 843-2473.

NOTES

INTRODUCTION

1. Williams and Williams, "An Oceanic Mass Migration of Land Birds," 173.

CHAPTER 1. FLIGHT ACROSS THE GULF

1. Quoted in Lowery, "Evidence of Trans-Gulf Migration," 191, from William Bullock, *Six Months' Residence and Travels in Mexico*, London: John Murray, 1824–l825.

2. Frazar, "Destruction of Birds by a Storm While Migrating," 251.

3. Lowery, "Evidence of Trans-Gulf Migration," 182.

4. George Williams, "Lowery on Trans-Gulf Migration," 228.

5. All quotes by Sidney Gauthreaux are from an interview, February 2, 2004.

6. Arvin, "Anchored in a River of Birds," 27.

7. Unless otherwise noted, all quotes by John C. Arvin are from an interview, May 9, 2003.

8. Arvin, "Anchored in a River of Birds," 29.

9. Ibid., 30.

10. Arvin, "Migration Spectacle in the Gulf."

CHAPTER 2. MAKING LANDFALL

1. Zoltán Németh, interview, October 6, 2004.

2. Quotes by Ashley Sutton were reconstructed based on conversations during field work at Johnsons Bayou that actually occurred on April 25, 2004, but that are typical for any given day of mist netting.

CHAPTER 3. THE THRUSH CHASERS

1. Graber, "Night Flight with a Thrush," 371.

2. Quotes by Richard Graber in this paragraph and next are from Graber, "Night Flight with a Thrush," 371.

3. Ibid., 373.

4. Ibid., 374.

5. Unless otherwise noted, all quotes by Martin Wikelski are from an interview, June 2, 2004.

6. Princeton University press release, "Sixth Sense: Study Shows How Migrating Birds Navigate."

7. Bill Cochran, interview, June 23, 2004.

CHAPTER 4. WITNESSING THE SPRING SPECTACLE

1. Fitzpatrick, "The View from Sapsucker Woods," 2.

CHAPTER 5. SUMMER SPLENDOR

1. Kroodsma, "Vocal Behavior," 7–2.

2. Darwin, *The Descent of Man, and Selection in Relation to Sex,* 39.

CHAPTER 6. WARBLERS AND WOODLAND INTRICACIES

1. All quotes from Richard T. Holmes were from an interview on June 27, 2003, at the Hubbard Brook Experimental Forest.

CHAPTER 7. HOW TO FIND AND MONITOR BIRD NESTS
1. Jim Berry, interview, January 28, 2004.

CHAPTER 8. FIVE THOUSAND MILES SOUTH BY NIGHT
1. Gosse, *The Birds of Jamaica*, 230.

CHAPTER 9. LOOKING AND LISTENING FOR AUTUMN MIGRATION
1. Bill Evans, interview, March 30, 2005.

CHAPTER 10. BIRDS OF TWO WORLDS
1. Jim Chace, quoted by Kent McFarland, interview, March 11, 2005.

CHAPTER 11. BIRDS OF NORTHERN WINTERS
The "lesser redpolls" that Thoreau mentions are now recognized as common redpolls, *Carduelis flammea*.
1. Thoreau, *Thoreau's Bird-Lore*, 419–20.
2. Ibid., 278–79, 284–85.
3. Ibid., 420.

EPILOGUE
1. John Deere's 1837 self-scouring, steel-bladed plow enabled settlers to plow the prairie faster than ever before (Carey, "Little Habitat on the Prairie"). For information on the transformation of tallgrass prairie to agricultural fields, see the National Park Service Web site at http://www.nps.gov/tapr/pphtml/nature.html.

2. Terborgh, *Where Have All the Birds Gone?* xiv, 8, 152.

3. Di Giacomo et al., "Status and Conservation of the Bobolink (*Dolichonyx oryzivorus*) in Argentina."

4. A study led by Eric Bollinger in New York found that 51 percent of bobolink eggs and nestlings were destroyed when a hayfield was mowed; subsequent nest abandonment, nest predation, raking, and baling increased mortality to 94 percent. More

than 50 percent of fledglings were estimated to have been killed. In the United States and Canada, bobolinks declined by 3 percent per year during 1980–2004, according to the Breeding Bird Survey. (Line, "A Summer Without Bobolinks," 10–34.)

5. Terborgh, *Where Have All the Birds Gone?* 48.

6. In the thirty-seven acre University of Delaware Woods in Newark, Delaware, parasitism by cowbirds was 0–5 percent, ten times lower than in a semiwooded residential area with lawns and wide streets and in fourteen suburban woods that were each less than six acres in size. Roth et al., "Wood Thrush (*Hylocichla mustelina*)."

7. Fitzpatrick, "Bird Conservation," 10–73, 10–74.

8. Numbers of birds killed by cats are difficult to estimate but John Coleman, Stanley Temple, and S. R. Craven calculated that rural free-roaming cats kill between 8 million and 217 million birds a year in Wisconsin alone ("Cats and Wildlife: A Conservation Dilemma"). In the United States, window collisions are estimated to kill between 97.6 and 975.6 million birds each year, according to Klem, "Collisions Between Birds and Windows: Mortality and Prevention."

9. Shire, Brown, and Winegard. "Communication Towers: A Deadly Hazard to Birds."

10. U.S. Fish and Wildlife Service, memo, "Service Guidance on the Siting, Construction, Operation and Decommissioning of Communications Towers."

11. American Bird Conservancy, http://www.abcbirds.org/policy/towerkill.htm.

12. Iñigo-Elias et al., "The Danger of Beauty."

13. Painted buntings declined by 1.9 percent per year between 1966 and 2004, according to the Breeding Bird Survey.

14. Di Giacomo et al., "Status and Conservation of the Bobolink (*Dolichonyx oryzivorus*) in Argentina."

15. Eduardo Iñigo-Elias, personal communication.

16. In the United States, where regulations are relatively stringent, some forty legally allowable pesticide ingredients are known

to cause bird die-offs. Pimentel et al., "Environmental and Economic Costs of Pesticide Use."

17. Rich et al., *Partners in Flight North American Landbird Conservation Plan.*

18. Gauthreaux, "Neotropical Migrants and the Gulf of Mexico," 47.

19. Total population estimates from Rich et al., *Partners in Flight North American Landbird Conservation Plan.* Loss percentages from Breeding Bird Survey.

20. Mayfield, "Kirtland's Warbler."

21. Fitzpatrick, "Bird Conservation," 10-29–10-30.

22. Blancher and Wells, "The Boreal Forest Region: North America's Bird Nursery."

23. Van der Voort and Greenberg, "Why Migratory Birds Are Crazy for Coffee."

24. Price and Glick, "The Birdwatcher's Guide to Global Warming."

25. Wikelski et al., "Costs of Migration in Free-Flying Songbirds," 704.

26. Martin E. Wikelski, personal communication, June 2, 2004.

27. Statistics on bird watchers and economic impact from "Birding in the United States: A Demographic and Economic Analysis." U.S. Fish and Wildlife Service report 2001-1.

28. Thoreau, *Thoreau's Bird-Lore,* 245–46.

BIBLIOGRAPHY

Able, K. P. "Birds on the Move: Flight and Migration." In *Handbook of Bird Biology*, ed. S. Podulka, R. W. Rohrbaugh Jr., and R. Bonney. 2nd ed. Ithaca, N.Y.: Cornell Laboratory of Ornithology, 2004.

———. "Migration Biology for Birders." *Birding*, April 1991.

"Acid Rain Research at the HBEF." http://www.hubbardbrook .org/education/Introduction/Intro13.htm.

Adkisson, C. S. "Red Crossbill (*Loxia curvirostra*)." In *The Birds of North America*. Ed. A. Poole and F. Gill. Philadelphia, Pa., and Washington, D.C.: The Academy of Natural Sciences and The American Ornithologists' Union, 1996.

Arvin, John C. "Anchored in a River of Birds." *Texas Birds* 3 (2001): 27–30.

———. "Migration Spectacle in the Gulf." *LOS News* 181 (1998), http://losbird.org/los_news_181_98july.htm.

Barnes, B. "Valley Wild: Tips for Birding the Southern Sierra Nevada and Kern River Valley and Annotated Checklist." 2005. http://www.valleywild.org/krvbirdtips.htm

Beason, R. C. "Use of an Inclination Compass During Migratory Orientation by the Bobolink (*Doliconyx oryzivorus*). *Ethology* 81 (1989): 291–99.

———. "You Can Get There From Here: Responses to Simulated Magnetic Equator Crossing by the Bobolink (*Dolichonyx oryzivorus*)." *Ethology* 91 (1992): 75–80.

Beason, R. C., N. Dussourd, and M. E. Deutschlander. "Behavioural Evidence for the Use of Magnetic Material in Magnetoreception by a Migratory Bird." *Journal of Experimental Biology* 198 (1995): 141–46.

Beason R. C., and J. E. Nichols. "Magnetic Orientation and Magnetically Sensitive Material in a Transequatorial Migratory Bird." *Nature* 309 (1984): 151–53.

Beason, R. C., and P. Semm. "Does the Avian Ophthalmic Nerve Carry Magnetic Navigational Information?" *Journal of Experimental Biology* 199 (1996): 1241–44.

———. "Magnetic Responses of the Trigeminal Nerve System of the Bobolink (*Dolichonyx oryzivorus*)." *Neuroscience Letters* 80 (1987): 229–34.

Benkman, C. W. "Adaptation to Single Resources and the Evolution of Crossbill (*Loxia*) Diversity." *Ecological Monographs* 63 (1993): 305–25.

Bent, A. C. "Bobolink." In *Life Histories of North American Blackbirds, Orioles, Tanagers, and Allies. United States National Museum Bulletin* 211 (1958): 28–52.

Berger, C. "Superflight." *National Wildlife* December/January 1999: 40–47.

Berry, J. "Finding Nests." *Birding,* April 1993: 110–117.

"Bicknell's Thrush: A Species Is Born." http://www.ns.ec.gc.ca/wildlife/bicknells_thrush/e/species_is_born.html.

"A Bird's-Eye View of Magnetic Earth," Virginia Tech press release, May 13, 2004.

Blancher, Peter, and Jeffrey W. Wells. "The Boreal Forest Region: North America's Bird Nursery." Report for Bird Studies Canada, the Boreal Songbird Initiative, and the Canadian Boreal Initiative, 2005.

Blom, E. A. T. "Fall Warblers: No Confusion." *Bird Watcher's Digest* 2005. http://www.birdwatchersdigest.com.

Blum, J. D., E. H. Taliaferro, and R. T. Holmes. "Determining the Sources of Calcium for Migratory Songbirds Using Stable Strontium Isotopes." *Oecologia* 126 (2001): 569–74.

Bock, C. E., and L. W. Lepthien. "Synchronous Eruptions of Boreal Seed-Eating Birds." *American Naturalist* 110 (1976): 559–71.

Bonta, M. "Audubon at Oakley." *Living Bird*, Winter 2002: 10–16.

Brown, C. R. "Purple Martin (*Progne subis*)." In *The Birds of North America*. Ed. A. Poole and F. Gill. Philadelphia, Pa., and Washington, D.C.: The Academy of Natural Sciences and The American Ornithologists' Union, 1997.

Brown, C. R., and M. B. Brown. *Coloniality in the Cliff Swallow*. Chicago: The University of Chicago Press, 1996.

Brown, R. M., S. Buff, T. Gallagher, and K. Streiffert. *Where the Birds Are: One Hundred Best Birdwatching Spots in North America*. New York: Doring Kindersley, 2001.

Burton, R. *Bird Migration: An Illustrated Account*. New York: Facts on File, 1992.

Buss, I. O. "Bird Detection by Radar." *Auk* 63 (1946): 315–18.

Carey, John. "Little Habitat on the Prairie." *National Wildlife*, June/July 2000: 52–59.

Chamberlain, C. P., et al. "The Use of Isotope Tracers for Identifying Populations of Migratory Birds." *Oecologia* 109 (1997): 132–41.

Cochran, W. W. "Orientation and Other Migratory Behaviours of a Swainson's Thrush Followed for 1,500 km." *Animal Behaviour* 35 (1987): 927–29.

Cochran, W. W., R. R. Graber, and G. G. Montgomery. "Migratory flights of *Hylocichla* Thrushes." *Living Bird* 6 (1967): 213–25.

Cochran, W. W., H. Mouritsen, and M. Wikelski. "Migrating Songbirds Recalibrate Their Magnetic Compass Daily from Twilight Cues." *Science* 304 (2004): 405–8.

Cochran, W. W. and M. Wikelski. "Individual Migratory Tactics of New World *Catharus* Thrushes: Current Knowledge and Future Tracking Options from Space." In *Birds of Two*

Worlds. Ed. P. Marra and R. Greenberg. Baltimore: Johns Hopkins Press, 2005.

Coleman, John S., Stanley A. Temple, and S. R. Craven. "Cats and Wildlife: A Conservation Dilemma." Madison, Wis.: U.S. Fish and Wildlife Service and University of Wisconsin-Extension, 1997.

Columbus, C. *Journal of First Voyage to America, by Christopher Columbus.* New York: Albert and Charles Boni, 1924.

Cooke, W. W. "Routes of Bird Migration." *Auk* 22 (1905): 1–11.

Darwin, Charles. *The Descent of Man, and Selection in Relation to Sex.* Princeton, N.J.: Princeton University Press, 1981.

Dietsch, Thomas. "Tennessee Warbler: The Coffee Warbler." *Bird of the Month.* Smithsonian Migratory Bird Center Web site http://nationalzoo.si.edu/ConservationAndScience/Migratory Birds.

Di Giacomo, Adrián S., A. G. Di Giacomo, and J. R. Contreras. "Status and Conservation of the Bobolink (*Dolichonyx oryzivorus*) in Argentina." In *Bird Conservation Implementation and Integration in the Americas.* Ed. T. Rich and C. J. Ralph. n.p.: USDA Forest Service, Pacific Southwest Research Station, forthcoming.

Dunn, P. O., and D. Winkler. "Climate Change Has Affected the Breeding Date of Tree Swallows Throughout North America." *Proceedings of the Royal Society Biological Sciences Series B* 272 (2005): 665–70.

Elphick, J., ed. *Atlas of Bird Migration: Tracing the Great Journeys of the World's Birds.* New York: Random House, 1995.

Farm Service Agency Online. http://www.fsa.usda.gov.

Fitzpatrick, John W. "Bird Conservation." In *Handbook of Bird Biology.* Ed. S. Podulka, R. W. Rohrbaugh Jr., and R. Bonney. 2nd ed. Ithaca, N.Y.: Cornell Laboratory of Ornithology, 2004.

———. "The View from Sapsucker Woods." *Birdscope* 16 (2002): 2.

FLAP. http://www.flap.org.

Fowle, M. T., P. Kerlinger, and W. Conway. *The New York City Audubon Society Guide to Finding Birds in the Metropolitan Area*. Ithaca, N.Y.: Cornell University Press, 2001.

Frazar, Martin A. "Destruction of Birds by a Storm While Migrating." *Bulletin of the Nuttall Ornithological Club* 6 (1881): 250–52.

Gallagher, T. "A Little Night Music." *Living Bird,* Spring 1996, 10–14.

Gauthreaux, Sidney Jr. "Neotropical Migrants and the Gulf of Mexico: The View from Aloft." In *Gatherings of Angels*. Ed. K. P. Able. Ithaca, N.Y.: Comstock Books, 1999.

———. "A Radar and Direct Visual Study of Passerine Spring Migration in Southern Louisiana." *Auk* 88 (1971): 343–65.

Godard, R. "Long-term Memory of Individual Neighbours in a Migratory Songbird." *Nature* 1991, 228–29.

Gosse, Philip Henry. *The Birds of Jamaica*. London: John Van Voorst, Paternoster Row, 1857.

Graber, Richard. "Night Flight with a Thrush." *Audubon Magazine* 67 (1965): 368–74.

Great Backyard Bird Count Web site. "Ebb-and-Flow Range Changes in Carolina Wrens." http://www.birdsource.org/gbbc.

Greenberg, R. "Cape May Warbler: The Flying Tiger." *Bird of the Month*. Smithsonian Migratory Bird Center Web site http://nationalzoo.si.edu/ConservationAndScience/MigratoryBirds.

———. "Constant Density and Stable Territoriality in Some Tropical Insectivorous Birds." *Oecologia* 69 (1987): 618–25.

———. "The Hairy Trees of Mexico." *Spotlight on Birds*. Smithsonian Migratory Bird Center Web site http://nationalzoo.si.edu/ConservationAndScience/MigratoryBirds.

———. "White-eyed Vireo: Bird of Many Voices." *Bird of the Month*. Smithsonian Migratory Bird Center Web site http://nationalzoo.si.edu/ConservationAndScience/MigratoryBirds.

———. "Worm-eating Warbler: Acrobat of the Aerial Leaf Litter." *Bird of the Month*. Smithsonian Migratory Bird Center

Web site http://nationalzoo.si.edu/ConservationAndScience/ MigratoryBirds.

Greenberg, R., M. Foster, and L. Marquez-Valdelamar. "The Role of White-eyed Vireos in the Dispersal of *Bursera simaruba* Fruit." *Journal of Tropical Ecology* 11 (1995): 619–39.

Greenberg, R., C. Macias Caballero, and P. Bichier. "Defense of Homopteran Honeydew by Birds in the Mexican Highlands and other Warm Temperate Forests." *Oikos* 68 (1993): 519–24.

Greenberg, R., and J. Reaser. *Bring Back the Birds.* Mechanics-burg, Pa.: Stackpole Books, 1995.

Griffith, S. C., I. P. F. Owens, and K. A. Thurman. "Extra Pair Paternity in Birds: A Review of Interspecific Variation and Adaptive Function." *Molecular Ecology* 11 (2002): 2195–2212.

Hames, R. S., et al. "Adverse Effects of Acid Rain on the Distribution of the Wood Thrush *Hylocichla mustelina* in North America." *Proceedings of the National Academy of Sciences* 99 (2002): 11, 235–40.

Hamilton, W. J. III. "Bobolink Migratory Pathways and Their Experimental Analysis under Night Skies." *Auk* 79 (1962): 208–33.

———. "Does the Bobolink Navigate?" *Wilson Bulletin* 74 (1962): 357–66.

Hill, G. E. *A Red Bird in a Brown Bag.* New York: Oxford University Press, 2002.

Hobson, K. A., and L. I. Wassenaar. "Linking Breeding and Wintering Grounds of Neotropical Migrant Songbirds Using Stable Hydrogen Isotopic Analysis of Feathers." *Oecologia* 109 (1997): 142–48.

Hochachka, W., et al. "Irruptive Migration of Common Redpolls." *Condor* 101 (1999): 195–204.

Holmes, R. T., P. P. Marra, and T. W. Sherry. "Habitat-Specific Demography of Breeding Black-throated Blue Warblers (*Den-*

droica caerulescens): Implications for Population Dynamics." *Journal of Animal Ecology* 65 (1996): 183–85.

Holmes, R. T., and T. W. Sherry. "Thirty-Year Bird Population Trends in an Unfragmented Temperate Deciduous Forest: Importance of Habitat Change." *Auk* 118 (2001): 589–609.

Holmes, R. T., T. W. Sherry, P. P. Marra, and K. E. Petit. "Multiple Brooding and Productivity of a Neotropical Migrant, the Black-throated Blue Warbler (*Dendroica caerulescens*), in an Unfragmented Temperate Forest." *Auk* 109 (1992): 321–33.

Howard, R. D. "The Influence of Sexual Selection and Interspecific Competition on Mockingbird Song (*Mimus Polyglottos*)." *Evolution* 28 (1974): 428–38.

Howell, S. N. G. *A Bird-Finding Guide to Mexico.* Ithaca, N.Y.: Comstock, 1999.

Hunt, P. D., and D. J. Flaspohler. "Yellow-rumped Warbler (*Dendroica coronata*)." In *The Birds of North America.* Ed. A. Poole and F. Gill. Philadelphia, Pa.: The Birds of North America, 1988.

Iñigo-Elias, Eduardo, K. V. Rosenberg, and J. V. Wells. "The Danger of Beauty." *BirdScope* 16 (2002): 1.

Johnson, A. J., and D. Bonter. "Fire, Drought, Beetles, and Birds." *BirdScope* 18 (2004): 1, 6–7.

Johnson, L. S., and W. A. Searcy. "Female Attraction to Male Song in House Wrens (*Troglodytes aedon*)." *Behaviour* 133 (1996): 357–66.

Kammermeier, L. M., and W. M. Hochachka. "Project Feeder-Watch Annual Report 1998–99." *BirdScope* (1999): 1–6.

Kerlinger, P. *How Birds Migrate.* Mechanicsburg, Pa.: Stackpole Books, 1995.

King, B. "Roosts on Rigs." *Bird Watcher's Digest,* March/April 2001: 68–75.

Klem, D. Jr. "Collisions Between Birds and Windows: Mortality and Prevention." *Journal of Field Ornithology* 61 (1990): 120–28.

Knox, A. G., and P. E. Lowther. "Common Redpoll (*Carduelis flammea*)." In *The Birds of North America*. Ed. A. Poole and F. Gill. Philadelphia, Pa.: The Birds of North America, 2000.

Koenig, W. D. "Synchrony and Periodicity of Eruptions by Boreal Birds." *Condor* 103 (2001): 725–35.

Koenig, W. D., and J. M. Knops. "Seed-Crop Size and Eruptions of North American Boreal Seed-Eating Birds." *Journal of Animal Ecology* 70 (2001): 609–20.

Kroodsma, Donald E. "Vocal Behavior." In *Handbook of Bird Biology*. Ed. S. Podulka, R. W. Rohrbaugh Jr., and R. Bonney. 2nd ed. Ithaca, N.Y.: Cornell Laboratory of Ornithology, 2004.

Levy, S. "Navigating with a Built-in Compass." *National Wildlife Magazine* 37 (1999): 32–33.

Line, Les. "A Summer Without Bobolinks." In *Handbook of Bird Biology*. Ed. S. Podulka, R. W. Rohrbaugh Jr., and R. Bonney. 2nd ed. Ithaca, N.Y.: Cornell Laboratory of Ornithology, 2004.

Linville, S. U., and R. Breitwisch. "Carotenoid Availability and Plumage Coloration in a Wild Population of Northern Cardinals." *Auk* 114 (1997): 796–99.

Loria, D. E., and F. R. Moore. "Energy Demands of Migration on Red-eyed Vireos (*Vireo olivaceus*)." *Behavioral Ecology* 1 (1990): 24–35.

Lowe, G., and R. Greenberg. "Yellow-rumped Warbler: This Bird Is Bound to Berry." *Bird of the Month*. Smithsonian Migratory Bird Center Web site http://nationalzoo.si.edu/Conservation AndScience/MigratoryBirds.

Lowery, George H. Jr. "Evidence of Trans-Gulf Migration." *Auk* 63 (1946): 175–211.

Lowther, P. E., et al. "Yellow Warbler (*Dendroica petechia*)." In *The Birds of North America*. Ed. A. Poole. Ithaca, N.Y.: Cornell Laboratory of Ornithology; http://bna.birds.cornell.edu/BNA/account/Yellow_Warbler.

Lynch, J. F., E. S. Morton, and M. E. Van der Voort. "Habitat Segregation Between the Sexes of Wintering Hooded Warblers (*Wilsonia citrina*)." *Auk* 102 (1985): 714–21.

Lyon, B., and R. Montgomerie. "Snow Bunting and McKay's Bunting (*Plectrophenax nivalis* and *Plectrophenax hyperboreus*)." In *The Birds of North America*. Ed. A. Poole and F. Gill. Philadelphia, Pa., and Washington, D.C.: The Academy of Natural Sciences and The American Ornithologists' Union, 1995.

Marra, P. P. "American Redstart: The 'Christmas Bird.'" *Bird of the Month*. Smithsonian Migratory Bird Center Web site http://nationalzoo.si.edu/ConservationAndScience/Migratory Birds.

———. "The Role of Behavioral Dominance in Structuring Patterns of Habitat Occupancy in a Migrant Bird During the Nonbreeding Season." *Behavioral Ecology* 11 (2000): 299–308.

Marra, P. P., and R. L. Holberton. "Corticosterone Levels as Indicators of Habitat Quality: Effects of Habitat Segregation in a Migratory Bird During the Nonbreeding Season." *Oecologia* 116 (1998): 284–92.

Marra, P. P., and R. T. Holmes. "Avian Removal Experiments: Do They Test for Habitat Saturation or Female Availability?" *Ecology* 78 (1997): 947–52.

———. "Consequences of Dominance-Mediated Habitat Segregation in American Redstarts During the Nonbreeding Season." *Auk* 118 (2001): 92–104.

Martin, S. G., and T. A. Gavin. 1995. "Bobolink (*Dolichonyx oryzivorus*)." In *The Birds of North America*. Ed. A. Poole and F. Gill. Philadelphia, Pa., and Washington, D.C.: The Academy of Natural Sciences and The American Ornithologists' Union, 1995.

Mayfield, Harold F. "Kirtland's Warbler." In *The Birds of North America*. Ed. A. Poole, P. Stettenheim, and F. Gill. Philadelphia, Pa., and Washington, D.C.: The Academy of Natural Sciences and The American Ornithologists' Union, 1999.

McAtee, W. L. "Observations on the Shifting Range, Migration and Economic Value of the Bobolink." *Auk* 36 (1919): 430–31.

McCracken, J. "Welcome Back Wally and Sue." *Long Point Bird Observatory Newsletter*. 27, no. 2 (1997): 18.

McDonald, M. V. "Function of Song in Scott's Seaside Sparrow (*Ammodramus maritimus peninsulae*)." *Animal Behaviour* 38 (1989): 468–85.

Middleton, A. L. A. "American Goldfinch (*Carduelis tristis*)." In *The Birds of North America*. Ed. A. Poole and F. Gill. Philadelphia, Pa., and Washington, D.C.: The Academy of Natural Sciences and The American Ornithologists' Union, 1993.

Moore, F. R. "Neotropical Migrants and the Gulf of Mexico: The Cheniers of Louisiana and Stopover Ecology." In *Gatherings of Angels*. Ed. K. P. Able. Ithaca, N.Y.: Comstock Books, 1999.

Moore, F. R., and P. Kerlinger. "Stopover Fat Deposition by North American Wood-Warblers (Parulinae) Following Spring Migration over the Gulf of Mexico." *Oecologia* 74 (1987): 47–54.

Morton, E. S. "Habitat Segregation by Sex in the Hooded Warbler: Experiments on Proximate Causation and Discussion of Its Evolution." *American Naturalist* 135: 319–33.

Morton, E. S., J. F. Lynch, K. Young, and P. Mehlhop. "Do Male Hooded Warblers Exclude Females from Nonbreeding Territories in Tropical Forest?" *Auk* 104 (1987): 133–35.

Mouritsen, H., et al. "Cryptochromes and Neuronal-Activity Markers Colocalize in the Retina of Migratory Birds During Magnetic Orientation." *Proceedings of the National Academy of Sciences* 101 (2004): 14,294–99.

Murphy, M. T., et al. "Population Structure and Habitat Use by Overwintering Neotropical Migrants on a Remote Oceanic Island." *Biological Conservation* 102 (2001): 333–45.

National Park Service. "Tallgrass prairie." http://www.nps.gov/tapr/pphtml/nature.html.

Naugler, C. T. "American Tree Sparrow." In *The Birds of North America*. Ed. A. Poole, P. Stettenheim, and F. Gill. Philadelphia, Pa., and Washington, D.C.: The Academy of Natural Sciences and The American Ornithologists' Union, 1993.

Neighborhood Nestwatch. http://sio.si.edu/Nestwatch/default.cfm.

Nolan, V. Jr., et al. "Dark-eyed Junco (*Junco hyemalis*)." In *The Birds of North America*. Ed. A. Poole and F. Gill. Philadelphia: The Birds of North America, Inc., 2002.

Ornat, A. L., and R. Greenberg. "Sexual Segregation by Habitat in Migratory Warblers in Quintana Roo, Mexico." *Auk* 107 (1990): 539–43.

Ouellet, H. "Bicknell's Thrush: Taxonomic Status and Distribution." *Wilson Bulletin* 105 (1993): 545–72.

Perdeck, A. C. "Two Types of Orientation in Migrating Starlings (*Sturnus vulgaris L.*) and Chaffinches (*Fringilla coelebs L.*) as Revealed by Displacement Experiments." *Ardea* 46 (1958): 1–37.

Pettingill, O. S. "Winter of the Bobolink." *Audubon* 85 (1983): 102–9.

Pimentel, David, et al. "Environmental and Economic Costs of Pesticide Use." *Bioscience* 42 (1992): 750–60.

Place, A. R., and E. W. Stiles. "Living off the Wax of the Land: Bayberries and Yellow-rumped Warblers." *Auk* 109 (1992): 334–45.

Price, J., and P. Glick. "The Birdwatcher's Guide to Global Warming." Report for the National Wildlife Federation and American Bird Conservancy, 2002.

Rappole, J. H, and D. W. Warner. "Ecological Aspects of Migrant Bird Behavior in Vera Cruz, Mexico." In *Migrant Birds in the Neotropics*. Ed. A. Keast and E. S. Morton. Washington, D.C.: Smithsonian Institution Press, 1980.

Reitsma, L. R., R. T. Holmes, and T. W. Sherry. "Effects of Removal of Red Squirrels (*Tamiasciurus hudsonicus*) and Eastern Chipmunks (*Tamias striatus*) on Nest Predation in a Northern Hardwood Forest: An Artificial Nest Experiment." *Oikos* 57 (1990): 375–80.

Rich, Terry D., et al. *Partners in Flight North American Landbird Conservation Plan*. Ithaca, N.Y.: Cornell Laboratory of Ornithology, 2004.

Rimmer, C. C., et al. "Bicknell's Thrush (*Catharus bicknelli*)." In *The Birds of North America*. Ed. A. Poole and F. Gill. Philadelphia: The Birds of North America, 2001.

Rimmer, C. C., and K. P. McFarland. "Known Breeding and Wintering Sites of a Bicknell's Thrush." *Wilson Bulletin* 113 (2001): 234–36.

Rising, J. D., and N. J. Flood. "Baltimore Oriole (*Icterus galbula*)." In *The Birds of North America*. Ed. A. Poole and F. Gill. Philadelphia: The Birds of North America, 1998.

Ritz, T., et al. "Resonance Effects Indicate a Radical-Pair Mechanism for Avian Magnetic Compass." *Nature* 429 (2004): 177–80.

Rodenhouse, N. L. "Potential Effects of Climatic Change on a Neotropical Migrant Landbird." *Conservation Biology* 6 (1992): 263–72.

Rodenhouse, N. L., and R. T. Holmes. "Results of Experimental and Natural Food Reductions for Breeding Black-throated Blue Warblers." *Ecology* 73 (1992): 357–72.

Rodenhouse, N. L., T. W. Sherry, and R. T. Holmes. "Site-Dependent Regulation of Population Size: A New Synthesis." *Ecology* 78 (1997): 2,025–42.

Roth, Roland R., M. S. Johnson, and T. J. Underwood. "Wood Thrush (*Hylocichla mustelina*)." In *The Birds of North America*. Ed. A. Poole and F. Gill. Philadelphia, Pa., and Washington, D.C.: The Academy of Natural Sciences and The American Ornithologists' Union, 1996.

Rubenstein, D. R., et al. "Linking Breeding and Wintering Ranges of a Migratory Songbird Using Stable Isotopes." *Science* 295 (2002): 1,062–65.

Sallabanks, R., and F. C. James. "American Robin (*Turdus migratorius*)." In *The Birds of North America*. Ed. A. Poole and F. Gill. Philadelphia: The Birds of North America, 1999.

Schwartz, P. "The Northern Waterthrush in Venezuela." *Living Bird* 3 (1964): 169–84.

————. "Orientation Experiments with Northern Waterthrushes Wintering in Venezuela." *Proceedings of the Thirteenth International Ornithological Congress.* 1963. 481–84.

Shire, G. G., K. Brown, and G. Winegard. "Communication Towers: A Deadly Hazard to Birds." American Bird Conservancy, 2000. http://www.abcbirds.org.

Sibley, D. *The Birds of Cape May.* Bernardsville: New Jersey Audubon Society, 1997.

Sillett, T. S., and R. T. Holmes. "Variation in Survivorship of a Migratory Songbird Throughout Its Annual Cycle." *Journal of Animal Ecology* 71 (2002): 296–308.

"Sixth Sense: Study Shows How Migrating Birds Navigate." Princeton University press release, April 28, 2004.

Smith, S. M. "Black-capped Chickadee." In *The Birds of North America.* Ed. A. Poole, P. Stettenheim, and F. Gill. Philadelphia, Pa., and Washington, D.C.: The Academy of Natural Sciences and The American Ornithologists' Union, 1993.

"Southeastern Arizona Bird Observatory Guide to Birding Hot Spots." http://www.sabo.org.

Stoddard, P. K. "Vocal Recognition of Neighbors by Territorial Passerines." In *Ecology and Evolution of Acoustic Communication in Birds.* Ed. D. E. Kroodsma and E. H. Miller. Ithaca, N.Y.: Cornell University Press, 1996.

Strickland, D., and H. Ouellet. "Gray Jay." In *The Birds of North America.* Ed. A. Poole, P. Stettenheim, and F. Gill. Philadelphia, Pa., Washington, D.C.: The Academy of Natural Sciences and The American Ornithologists' Union, 1993.

Taliaferro, E. H., R. T. Holmes, and J. D. Blum. "Eggshell Characteristics and Calcium Demands of a Migratory Songbird Breeding in Two New England Forests." *Wilson Bulletin* 113 (2001): 94–100.

Tarvin, K. A., and G. E. Woolfenden. "Blue Jay (*Cyanocitta cristata*)." In *The Birds of North America.* Ed. A. Poole and F. Gill. Philadelphia: The Birds of North America, 1999.

Taylor, R. C. *A Birder's Guide to Southeastern Arizona*. Colorado Springs: American Birding Association, 1999.

Temple, S. A. "Dickcissel (*Spiza americana*)." In *The Birds of North America*. Ed. A. Poole and F. Gill. Philadelphia: The Birds of North America, 2002.

———. "Individuals, Populations, and Communities: The Ecology of Birds." In *Handbook of Bird Biology*. Ed. S. Podulka, R. W. Rohrbaugh Jr., and R. Bonney. 2nd ed. Ithaca, N.Y.: Cornell Laboratory of Ornithology, 2004.

Terborgh, John. *Where Have All the Birds Gone?* Princeton, N.J.: Princeton University Press, 1989.

Terres, J. K. *The Audubon Society Encyclopedia of North American Birds*. New York: Alfred A. Knopf, 1980.

Tessaglia-Hymes, C. "The Hawthorn Orchard." *The Cayuga Bird Club* Web site http://www.birds.cornell.edu/cayugabirdclub/hawthorn.html.

Thomas, R. J., et al. "Eye Size in Birds and the Timing of Song at Dawn." *Proceedings of the Royal Society of London* 269 (2002): 831–37.

Thoreau, Henry David. *Thoreau's Bird-Lore*. Ed. F. H. Allen. New York: Houghton Mifflin, 1925.

Tomback, D. F. "Clark's Nutcracker (*Nucifraga columbiana*)." In *The Birds of North America*. Ed. A. Poole and F. Gill. Philadelphia: The Birds of North America, 1998.

Troy, D. M. "Recaptures of Redpolls: Movements of an Irruptive Species." *Journal of Field Ornithology* 54 (1983): 146–51.

Van der Voort, M., and R. Greenberg. "Why Migratory Birds Are Crazy for Coffee." Smithsonian Migratory Bird Center fact sheet no. 1. http://nationalzoo.si.edu/ConservationAndScience/MigratoryBirds/Fact_Sheets/.

Weidensaul, S. *Living on the Wind*. New York: North Point Press, 1999.

Wheatley, N., and D. Brewer. *Where to Watch Birds in Central America and the Caribbean*. London: A&C Black, 2001.

White, M. *National Geographic Guide to Birdwatching Sites.* Washington, D.C.: National Geographic Society, 1999.

Wiedner, D. S., et al. "Visible Morning Flight of Neotropical Landbird Migrants at Cape May, New Jersey." *Auk* 109 (1992): 500–510.

Wikelski, M., et al. "Costs of Migration in Free-Flying Songbirds." *Nature* 423 (2003): 704.

Williams, George G. "Do Birds Cross the Gulf of Mexico in Spring?" *Auk* 62 (1945): 98–111.

———. "Lowery on Trans-Gulf Migration." *Auk* 64 (1946): 217–38.

Williams, Timothy C., and Janet M. Williams. "An Oceanic Mass Migration of Land Birds." *Scientific American* 239 (1978): 166–76.

Willis, E. O. "The Role of Migrant Birds at Swarms of Army Ants." *Living Bird* 5 (1966): 187–231.

Wilson, P. L., M. C. Towner, and S. L. Vehrencamp. "Survival and Song-Type Sharing in a Sedentary Subspecies of the Song Sparrow." *Condor* 102 (2000): 355–63.

Wiltschko, W., et al. "Red Light Disrupts Magnetic Orientation of Migratory Birds." *Nature* 364 (1993): 525–27.

Winkler, D. W. "Nests, Eggs, and Young: Breeding Biology of Birds." In *Handbook of Bird Biology.* Ed. S. Podulka, R. W. Rohrbaugh Jr., and R. Bonney. 2nd ed. Ithaca, N.Y.: Cornell Laboratory of Ornithology, 2004.

Witmer, M. C. "Bohemian Waxwing (*Bombycilla garrulus*)." In *The Birds of North America.* Ed. A. Poole and F. Gill. Philadelphia: The Birds of North America, 2002.

Yezerinac, S. M., and P. J. Weatherhead. "Extra-Pair Mating, Male Plumage Coloration and Sexual Selection in Yellow Warblers (*Dendroica petechia*)." *Proceedings of the Royal Society of London* 264 (1997): 527–32.

*I*NDEX